Creating Change to Improve Science and Mathematics Education

Chong Ho Yu · Hyun Seo Lee

Creating Change to Improve Science and Mathematics Education

Lessons from Hong Kong

Springer

Chong Ho Yu
Azusa Pacific University
Azusa, CA, USA

Hyun Seo Lee
Azusa Pacific University
Azusa, CA, USA

ISBN 978-981-15-3155-2 ISBN 978-981-15-3156-9 (eBook)
https://doi.org/10.1007/978-981-15-3156-9

This Springer imprint is published by the registered company Springer Nature Singapore Pte Ltd.
The registered company address is: 152 Beach Road, #21-01/04 Gateway East, Singapore 189721, Singapore

Preface

Are US students ready for the global competition, which is keen and even brutal? Governments of many major countries have treated education and the development of human resources as a priority (Robinson, 2015). However, international assessments, such as the Programme for International Student Assessment (PISA), the Programme for International Assessment of Adult Competencies (PIAAC), and Trends for International Math and Science Study (TIMSS), have consistently shown that students in the United States lag behind their international counterparts in areas of math, science, and reading. This is by no means a new phenomenon. By the early 1970s, US students ranked last among major developed countries in seven of nineteen academic achievement tests (Wildavsky, 2011). In 1983, the National Commission on Excellence in Education compiled a report entitled *A Nation at Risk*. According to the report, "If an unfriendly foreign power had attempted to impose on America the mediocre educational performance that exists today, we might well have viewed it as an act of war" (US Department of Education, 2018, p. 2).

One could ask, "Why do we need another book about improving science and math education?" There have been numerous projects and initiatives on this very subject since the 1980s. In response to the alleged crisis, during the 1980s there was a movement in education, calling America "back to the basics," in an attempt to boost the performance of US students. In 1989, the National Council of Teachers of Mathematics (NCTM, 2000) disseminated new standards in the hope of improving mathematics education from Kindergarten through 12th grade. NCTM's *Standards for School Mathematics: Visions for Implementation* report (1989) clearly stated "ambitious but achievable" goals of educating students to value, communicate, reason, and be confident in mathematics (Crosswhite, Dossey, & Frye, 1989, p. 521). Furthermore, in 1993, the American Association for the Advancement of Science (AAAS) called for reforming the education system by announcing *Science for All Americans and Benchmarks for Science Literacy* (AAAS, 1993). As a follow-up to this initiative, AAAS launched *Project 2061*, which is a long-term research and development scheme aiming to improve science education so that all Americans might become fluent in science, mathematics, and technology (AAAS, 2013). In 1996, the National Research Council (NRC) set a high bar of science education by issuing the National Science Education Standards (NRC, 1996). In 2002, President Bush signed the *No Child Left Behind (NCLB) Act*, employing a heavy-handed policy against substandard teachers and schools (US Department of Education, 2003). In 2009, the National Science Teachers Association (NSTA) developed a new national science framework through the Anchors project with the hope that US learners would succeed globally (NSTA, 2009; Yu & Wu, 2014). However, as will be discussed later, the authors of this book count themselves alongside many other educators, policy makers, and education commissioners who are afraid that this prediction three decades ago was only partially correct: the objectives were only ambitious, but not achievable.

One reason why we fear that such objectives are not achievable is that calling for a national policy, or setting up a national standard, stands in a sharp contrast to the US tradition of federalism and decentralization. The 2002 *No Child Left behind Act*, which was an updated version of the 1965 *Elementary and Secondary Education Act*, is a good example of an ambitious but non-achievable programme. Both NCLB and ESE shared the same goal: pushing American students to meet the highest standards. However, NCLB attempted to achieve this goal by heavy-handed standardization. Specifically, schools that failed to meet the standard for two consecutive years would not receive federal funding. States must set goals to show "Adequate Yearly Progress" (AYP). In reaction to the act, many states simply adjusted their standards and even manipulated data in order to fulfill the requirements (Wiseman, 2015). Due to widespread resentment and questionable results, the direction set by NCLB was reversed in 2015 by the *Every Student Succeeds Act*, which shifted the control of education policy from the federal government back to state governments (Heise, 2017).

Furthermore, some of those reports, like "A Nation at Risk", adopted a gloomy view of the US education system and appealed to a sentiment of crisis. For example, the National Mathematics Advisory Panel (NMAP, 2008) wrote

> This Panel, diverse in experience, expertise, and philosophy, agrees broadly that the delivery system in mathematics education—the system that translates mathematical knowledge into value and ability for the next generation—is *broken and must be fixed*. This is not a conclusion about teachers or school administrators, or textbooks or universities or any other single element of the system. It is about how the many parts do not now work together to achieve a result worthy of this country's values and ambitions (p. 10).

The authors of this book are skeptical of this type of strong rhetoric ("the system is … broken"). As a matter of fact, there have been many improvements in education since the report "A Nation at Risk" was released. For instance, during the 1990s, the Clinton administration allocated hundreds of millions of dollars in Title I funds to schools in impoverished communities, and the results were positive (Wiseman, 2015). In addition, after the education reform launched by the Obama administration, today more than 24 million students, including 4 million black students, 3.5 million Hispanic students, and 2.8 million students with disabilities, have access to higher standards of education (US Department of Education, 2019).

The US is not alone in facing this issue of academic performance. UK students also perform at a lower level when compared with their peers in Asian countries regarding math and science. UK educators refer to this as the "England-East Asia gap" in mathematics (Jerrim & Choi, 2014). In 2014, the UK did something that would have been unthinkable in the past: the British government sent a team to Shanghai to learn how mathematics was taught there, and then invited about 70 math teachers from Shanghai to teach in England. The core feature of the Shanghai teaching style is that it is content-oriented rather than learner-oriented. Specifically, every student, no matter what he or she takes, is required to comprehend the mathematical concept before being allowed to move on to the next topic. Undoubtedly, the Chinese teaching methods that emphasize discipline were on a head-to-head collision course with British values, which gear toward personal freedom (BBC, 2015). Interestingly, after completion of a pilot study, the United Kingdom decided to adopt the Shanghai math teaching methods in 8,000 primary schools, which amounts to almost 50% of all primary schools in the UK. Additionally, 35 math teaching centers were established throughout the country to promote the Chinese method (CRI Online, 2015) and the outcomes are under ongoing evaluation (Boylan, Maxwell, Wolstenholme, Jay, & Demack, 2018; Wang, Barmby, & Bolden, 2017).

"Think outside the box!" This is always easy to say but more difficult to do. The problem is that most people are not aware of the box that confines us. Rather, we perceive that the box is an anchor and a shelter. Simply put, we take things for granted and refuse to leave our comfort zone. Ego can be another major hindrance. For a long time, the East has been willing to learn from the West. It takes humility and courage for Western nations to adopt Eastern education models. The UK has made a bold step.

The premise of this book is simple. We need to search beyond our national borders for a solution. On the other side of the world, Hong Kong, one of the top performers in international assessments and academic contests, has made similar academic proposals to those of the US concerning the significance of STEM education among young students. Students in Hong Kong appear to excel in math and science, aggressively chasing after Shanghai and Singapore. The authors argue that the US can significantly learn from Hong Kong for a range of reasons, which will be elaborated throughout the book. If the United Kingdom chose to learn from Shanghai, we believe that the US can similarly learn from Hong Kong. In making this strong claim, the authors by no means romanticize all Hong Kong educational policies and practices, nor do we claim that all of these methods are directly transferable across cultures.

One of the major differences between the US and many other countries is that the US education system is much more decentralized. Specifically, the US Constitution does not have any mandate for nationwide education provision. Given this context, it is understandable that although some innovative pedagogies used by Hong Kong originate from the US, American educators face many challenges in implementation. As mentioned before, the US has a decentralized system, and therefore education practices vary from state to state, from school district to school district, and even from school to school. Additionally, due to the instability of funding, many practices cannot be carried out in the long term. Moreover, information is often not freely shared between different institutions, and as a result, good education practices are slow to be disseminated. In contrast, the city of Hong Kong has the luxury of utilizing a centralized system to implement and fine-tune innovative pedagogies (Kwan, 2018). However, the authors of this book do not view a decentralized system as inherently problematic. On the contrary, it is meritorious because a decentralized system is beneficial to promote democratic values, innovation, and individual freedom. It is not surprising that many Asian parents who dislike conformity to national goals and standards continue to send their children to America for schooling.

With that said, the authors of this book do not suggest the wholesale adoption of Hong Kong's educational practices on a nationwide scale, like the strategy that the UK took regarding Shanghai's approach. Rather, we must recognize the positive side of the decentralized system in the US and adapt the Hong Kong approach to fit this cultural context. Thus, we ask our readers to keep an open mind. It is up to our readers to evaluate which of these Hong Kong approaches may be applicable to their particular setting (state, school district, school, classroom, etc.). Finally, we wish to express our gratitude to those Hong Kong educators who were willing to share their resources and methods with us, as well as Dr. Frances Wu and Ms. Josephine Wong who provided valuable input into this book project. We also gratefully acknowledge the funding received toward this research project from the Faculty Research Council at Azusa Pacific University.

Azusa, USA

Chong Ho Yu
Hyun Seo Lee

References

American Association for the Advancement of Science. (1993). *Project 2061: Benchmarks for science literacy*. New York, NY: Oxford University Press.

American Association for the Advancement of Science. (2013). *Project 2061*. Retrieved from http://www.project2061.org/publications/bsl/.

Boylan, M., Maxwell, B., Wolstenholme, C., Jay, T., & Demack, S. (2018). The mathematics teacher exchange and 'mastery' in England: The evidence for the efficacy of component practices. *Education Sciences, 8*, 1–31. https://doi.org/10.3390/educsci8040202.

British Broadcasting System. (2015). *Would Chinese-style education work on British kids?* Retrieved from http://www.bbc.com/news/magazine-33735517.

CRI Online. (2015). *Chinese education has come: Are British kids tough enough?* Retrieved from http://jiaoxue.cri.cn/20160719/d7eb75bd-97f9-a3d4-343a-c4353f5bce2e.html.

Crosswhite, F. J., Dossey, J. A., & Frye, S. M. (1989). NCTM standards for school mathematics: Visions for implementation. *Journal for Research in Mathematics Education*, *20*(5), 513–522.

Heise, M. (2017). From no child left behind to every student succeeds: Back to the future for education federalism. *Columbia Law Review, 117,* 1859–1896.

Jerrim, J., & Choi, Á. (2014). The mathematics skills of school children: How does England compare to the high-performing East Asian jurisdictions? *Journal of Education Policy, 29*, 349–376.

Kwan, S. M. (2018). Report of policies for supporting front-line STEM teachers. *Hong Kong Science Teachers' Journal, 133*, 31–50.

National Council of Teachers of Mathematics (NCTM). (2000). *Principles and Standards for school mathematics*. Reston, VA: Author.

National Mathematics Advisory Panel. (2008). *Foundations for success: The final report of the national mathematics advisory panel*. Washington DC: US Department of Education.

National Research Council. (1996). *National science education standards*. Washington, DC: The National Academies Press. https://doi.org/10.17226/4962.

National Science Teachers Association. (2009). *Project: Science anchor*. Retrieved from http://science.nsta.org/enewsletter/anchors.pdf.

Robinson, K., & Aronica, L. (2015). *Creative schools: The grassroots revolution that's transforming education*. New York, NY: Viking.

US Department of Education. (2003). *No child left behind*. Retrieved from https://www2.ed.gov/nclb/landing.jhtml.

US Department of Education. (2018). *Archived information: A nation at risk*. Retrieved from https://www2.ed.gov/pubs/NatAtRisk/risk.html.

US Department of Education. (2019). *Progress in our schools*. Retrieved from https://www.ed.gov/k-12reforms.

Wang, Y., Barmby, P., & Bolden, D. (2017). Understanding linear function: A comparison of selected textbooks from England and Shanghai. *International Journal of Science and Mathematics Education, 15,* 131–153. https://doi.org/10.1007/s10763-015-9674-x.

Wildavsky, B. (2011). Relax, America. Chinese math whizzes and Indian engineers aren't stealing your kids' future. *Think Again: Education*. Retrieved from https://foreignpolicy.com/2011/02/21/think-again-education/.

Wiseman, A. (2015). *How the world learns: Comparative educational systems*. Chantilly, VA: The Teaching Company.

Yu, C. H., & Wu, S. F. (2014). *Teaching math and science as a value system: Lessons from Asia*. Paper presented at Asian Conference on Education, Osaka, Japan.

Contents

Chapter 1
Introduction

1.1 Why the Hong Kong Model?

As its title implies, this book focuses on lessons from Hong Kong for science and math education. Hong Kong is not the only high-performing East Asian school system that could potentially offer inspiration to the United States. As mentioned in the preface, the United Kingdom has adopted the Shanghai model. Furthermore, as will be discussed in the next section, Singapore is the top performer in several international student assessments. Thus, one logical reaction that might arise in a perceptive reader's mind is: Why Hong Kong? Why didn't the authors propose the Shanghai model or the Singapore model? The rationale for our choice is presented below.

Both Hong Kong and Singapore have high standards of teaching and learning. Reportedly, the educational system in Singapore has produced youth and young adults who achieve the highest standardized test scores in the world. Additionally, among well-known international assessments and international university rankings, Singapore outperforms Hong Kong. At first glance, it may appear Singapore would work even better than Hong Kong as a successful model for the US. The authors themselves pondered this issue at the beginning of the project, which led to preliminary research on several East Asian educational models, including Japan, China, South Korea, Singapore, and Hong Kong. Informed by both primary and secondary sources, the authors decided that Hong Kong's educational model is a more appropriate and plausible one to introduce to the US. The reasons for this choice are as follows.

One of the secrets to Singapore's success is rather simple: a competitive academic setting and driven students (Huan, See, Ang, & Har, 2008; Tan, 2011). Singaporean education is highly focused on academic and career development, rather than comprehensive personal growth. Gaining entrance to top-tier universities and landing high-end careers after graduation are of the utmost importance in Singapore's education system. Singaporean parents wish the best for their children in the given context, not only to survive but to excel in the academic arena. Parentocracy is a well-known term associated with Singapore, where one's educational goals are

C. H. Yu and H. S. Lee, *Creating Change to Improve Science and Mathematics Education*, https://doi.org/10.1007/978-981-15-3156-9_1

exclusively oriented toward fulfilling the wishes of his or her parents, and the quality of one's education is largely conditional upon the wealth of one's parents (Tan, 2017). Parents tend to walk their children along the so-called elite course. Children officially start their formal education at the age of seven, while in fact many begin as early as three. Private tutoring and home tuition are highly common in Singapore, with such additional and intensive academic support lasting from primary school to junior college.

In elementary and secondary education in Singapore, instructors are less avoidant of employing a military training mindset (Ngiam & Tung, 2016). Furthermore, instructors and institutions proceed with harsh punishments for students as a legal disciplinary method (Jyoti & Neetu, 2013). Stressing discipline and conformity is deeply seeded in Singapore classrooms, and the lectures are teacher-oriented (Sing & Khine, 2008). School regulations may involve having students line up for morning assemblies, recite Singapore's pledge, and sing the national anthem. Although these may not be directly linked to students' academic excellence, they contribute to maintaining the overall disciplinary atmosphere that can control their behaviors.

Parents, teachers, and private tutors hold high expectations for children and place a strong emphasis on test-taking results, as each partaker benefits from children's academic success (Cheo & Quah, 2005). For instance, the schools and private tutors can earn a good reputation in the community when their students are admitted to high-ranked domestic or international universities. In other words, Singaporean children are exposed to a decade-long barrage of academic and social pressure in a persistently continuous manner in their homes and schools (Cheow, 2019). A certain amount of stress aids in enhancing learning outcomes and drivenness can be a mark of high productivity (McGonigal, 2016). However, when enormous pressure to outperform one's competitors remains one of the most important drives of one's life, adverse effects to a student's understanding of himself or herself and the world can accrue (Ang & Huan, 2006; Fisher, 1994). Concerns over a lack of personal and socio-developmental aspects of children's growth have risen within Singapore. Furthermore, the elitism-focused academic trends of Singapore, which have been exacerbated by the rituals of private tutoring, are not in line with the US educational ethos and should not be implemented.

While traditionally Hong Kong has been traveling on a similar path to that of Singapore, it has shifted its educational focuses to suit its students' development better (Cheong, 2009; Forlin, 2010). Hong Kong's educational reform centers on developing a broad and balanced curriculum that can nurture an individual's whole-person development while retaining the importance of academic work. This main objective of Hong Kong's educational reformation shares some common ground with the scholastic philosophy of the US: education is understood not only to have an academic component but also to include a continual process of human development. Hong Kong's model, with its recent reformation, has found its own refined ways to form a good balance between traditional and modern goals. For instance, since the reform in 2008, the Hong Kong Educational Bureau has begun to de-emphasize the traditional school banding system, which classifies and allocates students in different schools based on student abilities. The motivation behind this reform, following the earlier one in 2000, was to abolish elitism and social stigma placed on young shoulders of Hong Kong students (Law, 2004; Pearson & Rao, 2006). The banding system is slowing dying, and educational institutions, parents, and students alike are looking forward to completely eliminating the system in the coming years. Another changing aspect is a student-oriented teaching style and classroom environment that encourages teacher and student dyads, openness, and creativity.

The last main point, creativity, is an indispensable component for acquiring new skills and developing a healthy personality. Although this factor will be discussed in a later session, it necessitates further articulation here as well. In a study comparing policies and strategies in developing creativity education in Asian Chinese societies, the data collected by Hui and Lau (2010) reveal that Hong Kong devoted more efforts to various aspects of creativity education than Mainland China and Singapore. To be more specific, when raters in this study reviewed educational documents related to creativity, Hong Kong received the highest score among the three Chinese societies in terms of identification of the component by the rater (see Table 1.1).

Interestingly, although both Singapore and Hong Kong were colonized and influenced by the United Kingdom, and both populations are mainly composed of Chinese people, their governing styles are vastly different from each other. While strong government intervention plays an important role in the Singaporean society, Hong Kong, on the contrary, embraces a positive nonintervention policy that minimizes the role of government in the free market (Wang, 2018). To explain this difference, Wong (2015) argued that because Singapore was bordered by two Muslim nations (Indonesia and Malaysia), and also faced the threat of the Chinese Communists, the Singaporean government had to tightly control the school system to contain influences from both the Western churches and the local Chinese. Unlike Singapore, after World War Two,

Table 1.1 Coverage of different aspects of creativity education in Asia

	Mainland China	Hong Kong	Singapore
1. Is there an individual policy on creativity education?			
2. Is creativity education part of another policy?	2	3	1
3. Does the policy discuss creativity in the curriculum?	1	3	1
4. Does the policy discuss creativity in extracurricular activities?		2	
5. Does the policy discuss creativity in the arts and cultural experiences?		2	
6. Does the policy discuss creativity in other learning experiences?		2	
7. Does the policy discuss creativity in testing?		2	
8. Is creativity education mandated?	1	2	
9. Are programs mandated with funding?	1		
10. Are programs not mandated but urged?	1	2	2

the government of Hong Kong was seldom challenged by hostile forces, except the Leftist rebellion in 1967. Thus, Hong Kong's systems, including its education system, could enjoy a high degree of uninterrupted freedom from strong government intervention. As mentioned in the preface, the structure of the US government is decentralized, entailing that the decentralized character of Hong Kong is much more compatible with the US.

Superiority in international assessments and university rankings can be a significant aspect of the worth of an educational model. However, the authors believe that a good educational model includes more than just perfect scores; it demands multifaceted adeptness and transferability of such advantages to other systems. In many ways, which will be discussed later throughout multiple chapters, Hong Kong's educational model will greatly benefit the current educational system in the United States.

1.2 Why Math and Science Education?

Why is math and science education important? One might list several noble goals: equipping future scientists with the tools to unveil the deeper secrets of the universe, or helping citizens think logically. One obvious answer is that regardless of whether we like it or not, people around the world are competing with each other for limited resources and opportunities in our high-tech economy. High-tech industries are crucial to many major national economies, as well as the global economy as a whole. Today, global competition is keen in various technological domains, such as artificial intelligence (AI), machine learning, big data analytics, 5G networks, etc. In February 2019, the US President issued an executive order to form a national initiative related to AI and machine learning. Before that time, 18 other nations had already treated AI research as a national priority. The US executive order stated, "Artificial Intelligence (AI) promises to drive growth of the United States economy, enhance our economic and national security, and improve our quality of life… The United States must train current and future generations of American workers with the skills to develop and apply AI technologies to prepare them for today's economy and jobs of the future" (Trump, 2019, Sect. 1).

This global race necessitates a well-trained workforce in science and technology, which is built upon effective science and math education. The Bureau of Labor Statistics (US Department of Labor, 2017) forecast that the fastest-growing jobs from 2016 to 2026 in the US would require a solid foundation in mathematics, science, and technology. These fields include solar photovoltaic technicians (105% growth), wind turbine service technicians (96%), statisticians (33%), software developers (30%), and mathematicians (29%). Additionally, electrical engineers are projected to grow 7%, as around 21,300 new electrical engineering jobs are expected by 2026. The overall employment rate of mathematicians and statisticians is projected to grow by 33% by 2026 (U.S Department of Labor, 2019). IBM predicts that by 2020, job openings related to data science will soar by 28% (Columbus, 2017). In 2016, after a thorough analysis of 54 million employee profiles across 350 industries with 15,000 job titles, *Money Magazine* and Payscale.com concluded that the top four most valuable skills are all related to data science (Renzulli, Weisser, & Leonhardt, 2016).

While the link between science education and technology-driven economy is obvious, the role of mathematics education in global technological competition needs further clarification. As has been mentioned, machine learning and big data analytics have become key technologies in the global economy, and these subject matters require a solid foundation in mathematics, such as matrix algebra, vector geometry, and probability (Deisenroth, Faisal, & Ong, 2019; Kepner & Jananthan, 2018). Furthermore, computer programming, one of the indispensable components of research and development in high-tech, requires mathematical and logical reasoning. Hong Kong educators are highly aware of the integrative nature of math, science, and technology. For example, biotechnology is one of the major emerging technologies that is revolutionizing the health care industry. Grobler (2018) said that we need big data analytics, which requires mathematics, to sequence DNA and to analyze gene functions. In addition, mathematics is also used to predict protein structures that would not have been possible without the availability of contemporary technology.

In spite of the high demand for jobs related to math and science, several international assessments—including Trends in International Mathematics and Science Study (TIMSS), the Programme for International Student Assessment (PISA), and the Programme for International Assessment of Adult Competencies (PICCA)—indicate that the US is at or below average in math and science performance, while other countries, especially Asian nations, rank among the top positions. In response to this performance gap, some authors (e.g., Emeagwali, 2010; Leshner, 2009) are concerned that US competitiveness in science, technology, engineering, and mathematics (STEM) is losing ground. On many occasions, President Obama called out the nation's need for more scientists and engineers in order to be competitive in the global arena, even introducing the phrase "Sputnik moment" to characterize the performance gap (Calmes, 2010; Kornblut & Wilson, 2011). His administration emphasized a strong causal link between education and national economic competitiveness while advocating for the American Recovery and Reinvestment Act (West, 2012).

One common response to the preceding point is that there is more to life than learning science and math; some people would rather choose human flourishing than seek to fill lucrative job slots and earn higher profits. Contrary to popular belief, the reformed Hong Kong secondary school curriculum is designed according to how the learning experience can enrich the student's life, not what manpower is demanded by the economy. Under the British system, before 1997, Hong Kong public exams were nothing more than screening tools for the universities that practiced elitism. The reformed curriculum in secondary schools was created before seeking endorsement from universities so that the curriculum could bring lifelong benefits for students, rather than serve as a stepping stone for academic and career advancement (OECD, 2011).

The authors by no means believe that STEM is more important than all other disciplines and aspects of life; rather, we believe that all disciplines and sub-disciplines in academia provide their unique contribution to human life and development, and interdisciplinary integration can offer a more holistic perspective for any subject matter under study. It is important to point out that Hong Kong does not have a vibrant high-tech industry, yet many Hong Kong teachers, parents, and students who were interviewed by the authors subscribe to the notion that logical reasoning and other scientific approaches can facilitate a good life, and that scientific modes of thinking are readily transferrable to other domains.

As a matter of fact, the demarcation between arts/humanities and science/engineering is a fairly recent phenomenon. During the Renaissance era, some well-known artists had an intense curiosity about various scientific fields. For example, Leonardo da Vinci studied a variety of scientific subjects, ranging from human anatomy to engineering. His knowledge of human anatomy contributed to his precise representation of human gestures and figures in drawings and paintings, such as the famous smile on the *Mona Lisa*. His exploration of optics also paved the way to the creation of the *Last Supper*, which illustrates how theatrical light could be well utilized in art (Isaacson, 2017). Another fruitful intersection of art and science can be found in music. According to German philosopher Gottfried Leibniz, “music is a secret exercise in arithmetic of the soul, unaware of its act of counting” (Cited in Mathematical Association of America, 2019, p. 18). Scientific knowledge of sound, such as the theory of harmonic progression and partial differential equations, has been found to be instrumental to musical composition and performance (Kung, 2013). According to Weiss, Murray, and Cyrus (2010), math and music have intertwined for many centuries. For example, the ancient Greeks infused music with geometrical structures, whereas J. S. Bach used a computational approach to compose music. In the modern era, avant-garde composers also explicitly borrowed ideas from probability and other mathematical fields (Kung, 2013). Further, Root-Bernstein and Root-Bernstein (2004) argued that both artists and scientists utilize interdisciplinary mental toolkits, such as observing, imaging, abstracting, patterning, empathizing, and so on, to nurture innovations. Therefore, many artists are indeed “scientific artists” while many scientists are “artistic scientists”.

The preceding brief discussion only begins to cover the interaction between art and science. The list can go on and on. For example, social scientists can benefit from training in statistical methods, whereas philosophers who specialize in epistemology can extract useful references from logic and the history of science. Even if we put aside academics, scientific knowledge can still enable non-academicians to make informed decisions for enhancing wellbeing. For example, while facing the crisis of climate change, we need certain scientific information in order to determine how to live a sustainable lifestyle.

Additionally, longitudinal research in the UK reveals that people with low math skills are more likely to be at risk of poor mental and physical health. To be specific, men with poor numeracy were more at risk of depression, were disinterested in politics, and also were more likely to have been suspended by the school, arrested, or cautioned by the police. Women with poor numeracy had similar problems and were also more likely to report poor physical health, to have low self-esteem, and to perceive that they lacked control over their lives (Parsons & Bynner, 2005). Moreover, although creativity education can be implemented in virtually any subject matter, scientific inquiry can no doubt facilitate curiosity, imagination, creativity, and openness. Prior research in human development and personality indicates that openness is linked to better physical health (Bjorklund, 2015).

Hence, it is the conviction of the authors that the US can still benefit from Hong Kong's education model so that the country can grow stronger across all disciplines, including the arts and humanities, as well as improving the mental and physical well-being of its citizens through math and science education. This is important because, in spite of the attractive values of math and science, US learners seem to have fallen behind their international peers in these fields.

1.3 Summary of International Assessments

This section offers a summary of the results of recent international assessments. The objective of this summary is to show quantitatively which nations are doing better in which areas. While taking the margins of error into account, readers should view these numbers as a reference only. It is not advisable to treat these statistics as absolute evidence for supporting the rhetoric that the US system is broken or that the entire nation is in crisis. Wiseman (2015) points out the disadvantages of international benchmarking: "Raising standards and expanding the curriculum are typical crisis responses, but a culture of crisis puts us at a disadvantage when it comes to using international comparisons in education… The reason is that crisis creates panic, and panic leads to reform and change for the sake of reform and change—not targeted and real reform that is aimed at improvement" (pp. 17–18). On the other hand, Wiseman still acknowledges the usefulness of international benchmarking: "Achievement rankings do not paint an accurate picture of education in the United States or abroad, because they do not account for context. But this also does not mean policymakers or educators should ignore the rankings or toss out the international studies" (p. 39).

1.3.1 TIMSS

Since 1995, Trends in International Mathematics and Science Study (TIMSS) has been evaluating students in Grade 4 and Grade 8 every four years. In order to obtain a representative sample from each participating nation, TIMSS asks each country to include at least 95% of the nationally desired populations of students. The TIMSS sample design adopted a two-stage stratified cluster sampling scheme. At the first stage, schools are sampled with probability proportional to size. At the second stage,

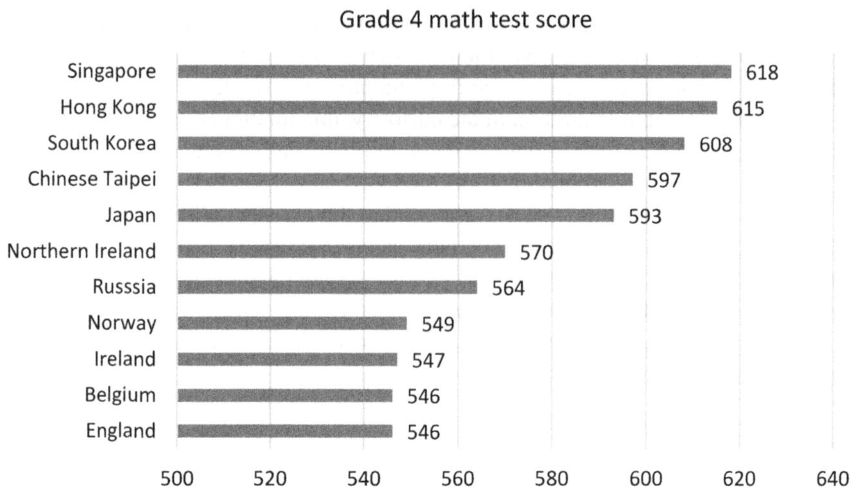

Fig. 1.1 Top performers of Grade 4 math test in 2015 TIMSS

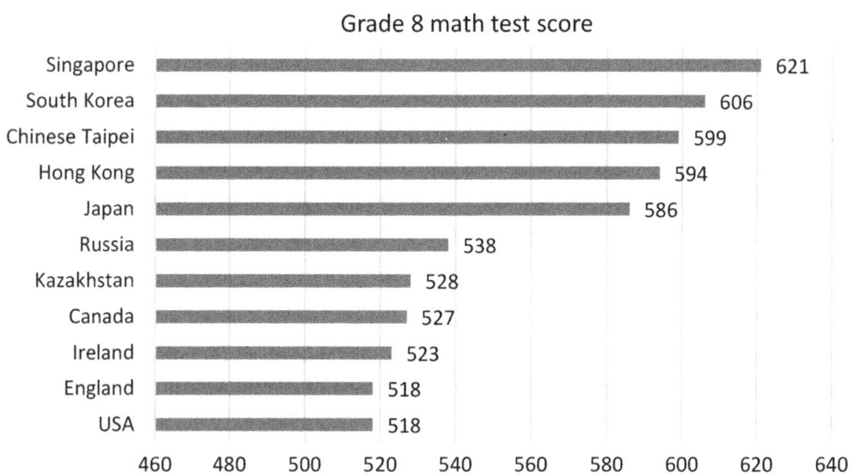

Fig. 1.2 Top performers of Grade 8 math test in 2015 TIMSS

one or more intact classes of students from the target grades are drawn (Joncas, 2008).

US students were outperformed by their international peers in past TIMSS tests. In 2015, 10 countries out of 48 total obtained statistically higher mean scores than the US in Grade 4 math test. In the science test, seven countries showed higher mean scores than the US. Among Grade 8 participants, 7 out of 37 countries outperformed the US in both math and science. Notably, most of the top performers in Grade 4 who took the 2015 TIMSS math test were from Asian countries (see Figs. 1.1, 1.2, 1.3 and 1.4).

In general, Asian students are believed to be good at memorizing and rote learning, rather than problem-solving. Contrary to this common belief, the TIMSS results show Asian countries and regions had the highest percentages of students reaching the Advanced International Benchmark for science, which indicates competency in the most complex topics and reasoning skills. For example, TIMSS 2007 data indicated that regarding Grade 4, Singapore and Chinese Taipei had 36 and 19% of their

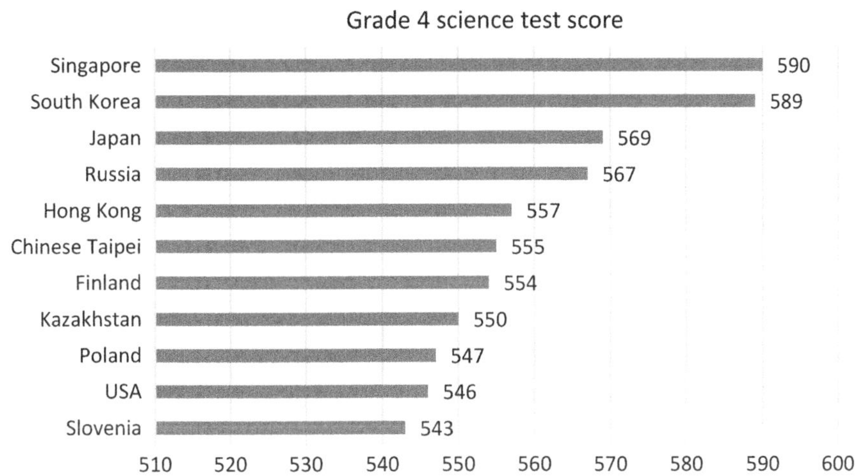

Fig. 1.3 Top performers of Grade 4 science test in 2015 TIMSS

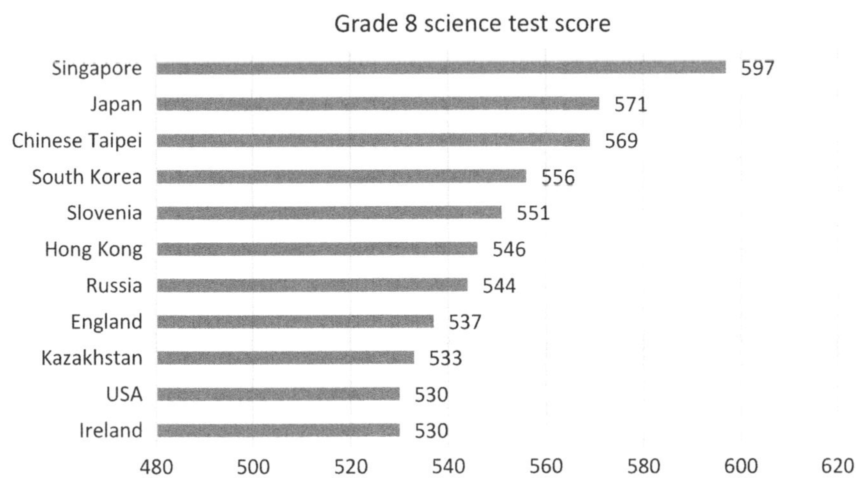

Fig. 1.4 Top performers of Grade 8 science test in 2015 TIMSS

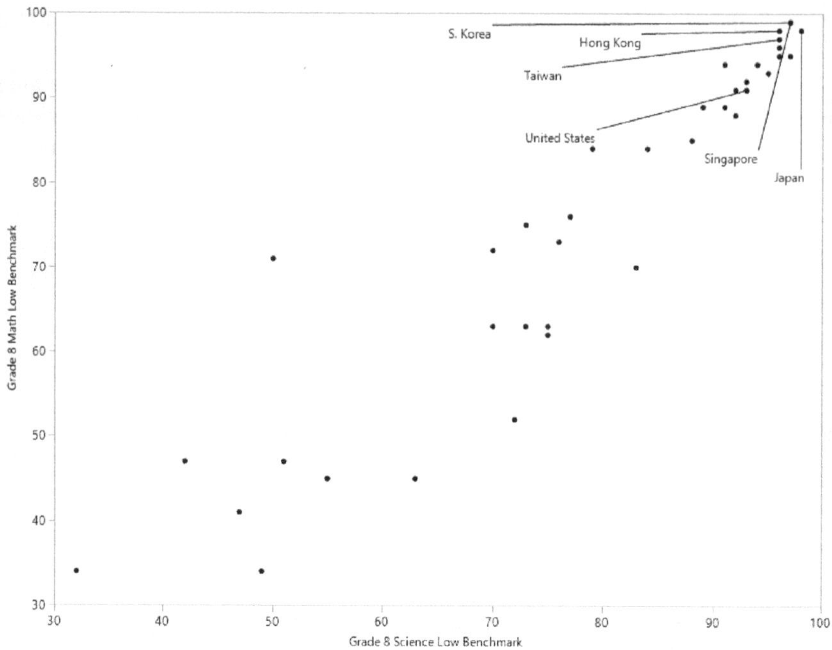

Fig. 1.5 TIMSS 2015 science and math benchmark: minimal performance gap between the US and East Asia among low performers

students, respectively, achieving at or above the benchmark. At Grade 8, Singapore and Taiwan had 32 and 25% of their students, respectively, meeting at or exceeding the benchmark. The median of developed countries reaching this benchmark was 7% at Grade 4 and 3% at Grade 8. More recently, in the 2015 TIMSS Advanced International Benchmark for science and math, the overall performance gap between the US and Asian countries was minimal (see Fig. 1.5). But the gap among mid-level and top performers between the US and Asia became even wider (see Figs. 1.6 and 1.7).

1.3.2 PISA

Since 2000, PISA administered by Organization for Economic Cooperation and Development (OECD) has delivered tests to 15-year-old students recruited from their member and nonmember countries every three years. Unlike TIMSS exams that are mostly content-based, PISA exams are scenario-based. Specifically, TIMSS exams cover the topics taught in school curricula, while PISA exams emphasize deduction, inference, and application in ways that may not directly be found in textbooks, such as finding the best route on a subway map in terms of time spent in traveling and costs, assigning participants to dorm rooms in a summer camp given

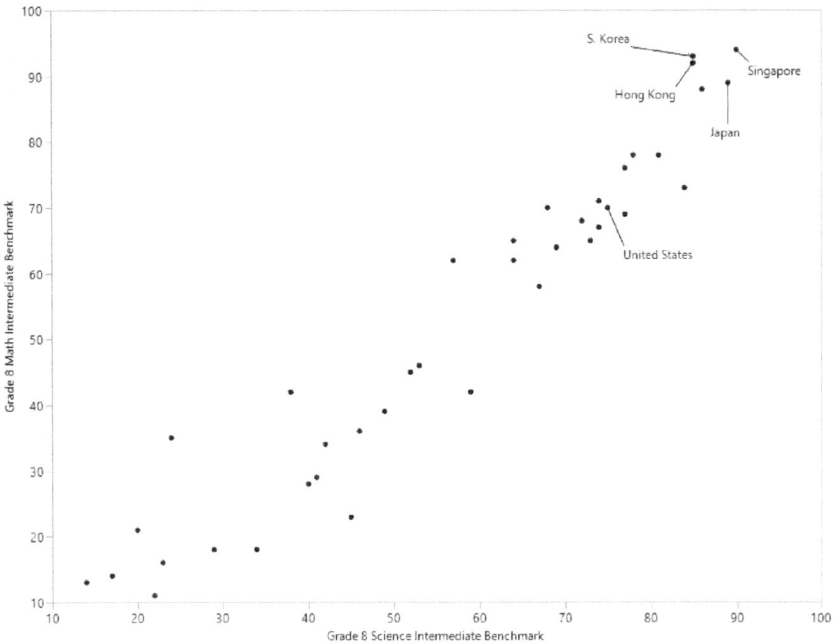

Fig. 1.6 TIMSS 2015 science and math benchmark: wider performance gap between the US and East Asia among low performers

certain rules and regulations, troubleshooting a malfunctioning irrigation system (Csapó & Funke, 2017). OECD countries realize that these days, most manual or routine cognitive tasks can be performed either by cheaper laborers or by computers; thus they expect that the demand for these types of jobs will continue to migrate from OECD countries to developing nations. In order for OECD citizens to fully participate in the process of globalization, there are some critical skills that emerging adults must possess: advanced problem-solving skills that go beyond just following protocols and fluent communication skills to illustrate complex scientific ideas in a user-friendly fashion. PISA has acknowledged a need for these skills and based on its educational philosophy. PISA items are designed to test students' ability to apply knowledge to various contexts rather than simply recalling names and terms (OECD, 2007).

Like TIMSS, PISA (OECD, 2013b) also employed stratified sampling, also known as multi-stage sampling, to cover the entire student population in the participating countries. The sampling process consisted of two stages. During the first stage, schools were selected, and at the second stage, students were sampled within the schools. This required the selection of at least 4,500 students and 150 schools in each participating nation.

Similar to TIMSS results, the US performance ranking was low in many PISA tests. In 2015, the US ranked 38th in math and 24th in science, out of 71 total

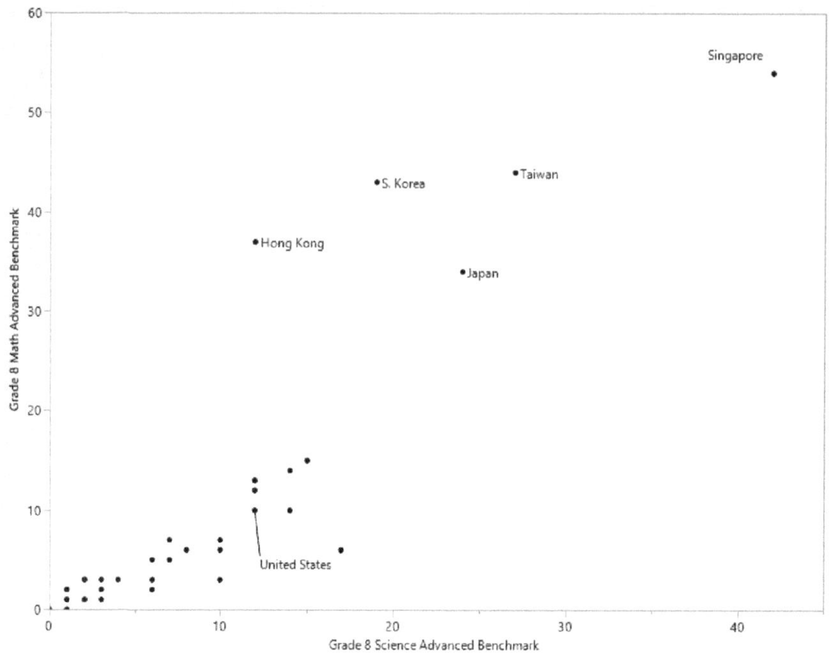

Fig. 1.7 TIMSS 2015 science and math benchmark: widest performance gap between the US and East Asia among advanced performers

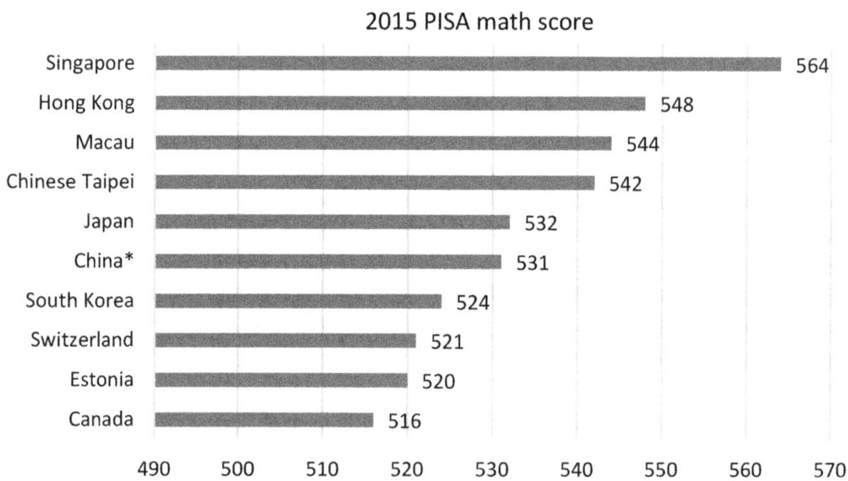

Fig. 1.8 Top performers of math test in 2015 PISA (*only four cities [Beijing, Shanghai, Jiangsu, and Guangdong] from China participated in 2015 PISA)

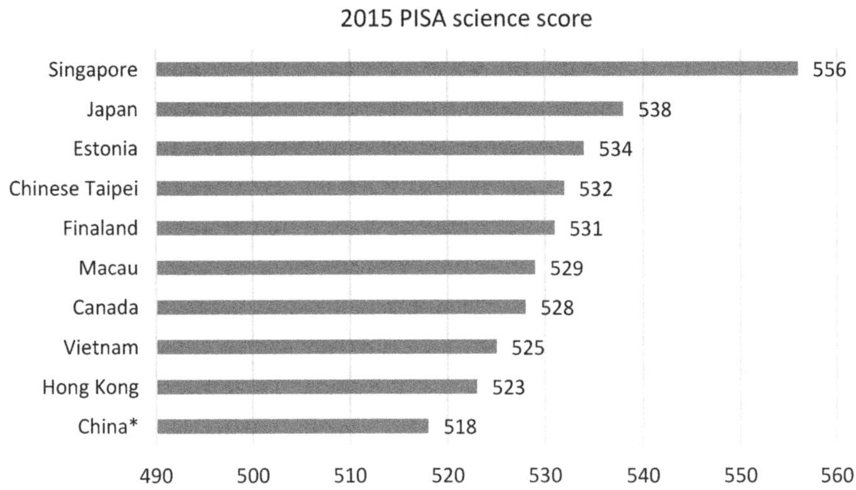

Fig. 1.9 Top performers of math test in 2015 PISA (*only four cities [Beijing, Shanghai, Jiangsu, and Guangdong] from China participated in 2015 PISA)

countries. Among the 35 OECD members, the US ranked 30th in math and 19th in science. Once again, the top performers of PISA were Asian countries and regions, as indicated by Figs. 1.8 and 1.9.

1.3.3 PIAAC

The Programme for International Assessment of Adult Competencies (PIAAC) conducted by OECD is an extension of PISA. The target population of the former consists of adults, whereas the population of the latter includes 15-year-old students. PIAAC research is sponsored by the US Department of Education's National Center for Education Statistics (NCES) and has been used to evaluate global trends regarding adult learning, as well as the current status of adult learning in the US (Linton, 2010). This is valuable information for educational leaders because it offers global trends, as well as national ability. With this information they can make educated and informed decisions concerning adult education in the US.

The 2016 PIAAC data demonstrated that American adults are far behind their international counterparts in all three test categories (literacy, numeracy, and problem-solving in technology-rich environments). The OECD conducted two rounds of data collection. Round 1 data were collected from 24 countries from 2011 to 2012. In 2013, the Round 1 report was released, showing that US youth aged 16–24 placed at the bottom in categories of numeracy and problem-solving in technology-rich environments. At this point, it comes as a relief that they ranked 17th in literacy. When all age groups were put together, the US ranked 16th in literacy, 21st in numeracy, and 14th in problem-solving (OECD, 2012, 2013a).

Nine additional countries participated in the second round of data collection during 2014–15, and the report was released in June 2016. After adding the data of nine more countries, the new analysis revealed that the US ranked 18th in literacy, 27th in numeracy, and 16th in problem-solving. Certain findings demand further reflections and discussions among US educators. For example, younger Singaporeans (aged 16–24) outperformed the older generation (aged 55–64) in literacy. However, the opposite trend obtained in the US, where older US adults performed with better literacy skills than young US adults. A similar pattern was observed in the technology-based problem-solving category. Many older US adults reached higher skill levels in problem-solving, and the US was second from the top, just behind New Zealand. However, the problem-solving skills of US young adults were among the bottom six nations (OECD, 2016). Recent research, which has utilized cohorts into an analysis, indicated that there is a strong relationship ($r = 0.7$) between PISA trends and the PIAAC performance among 20 countries. This could imply that the high quality of schooling for young learners has lasting effects on adult skills (Gustafsson, 2016). Conversely, a weak foundation of education will place individuals in a vulnerable position, where disadvantages are more likely to persist, and even to be exacerbated, throughout their lives (Yu, Lee, Lara, & Gan, 2019).

1.3.4 Implications

Hunt and Wittmann (2008) found that the Program for International Student Assessment (PISA) test scores provide an even better prediction of national wealth than TIMSS. Similarly, based on international test scores, including PISA, Hanushek and Woessmann (2008) concluded that the cognitive skills of the population are strongly associated with individual financial earnings and economic growth. After using the 1960 GDP per capita as the baseline and controlling for the average years of schooling in the same year, another interesting correlation was found: increasing test scores by one standard deviation was associated with increasing the annual GDP growth rates by almost two percent. Based on this relationship, if the performance of US students in math and science tests were to reach the top level, it would have increased the US growth rate by over a full percentage point. The link between cognitive skills measured by international tests and economic growth was also supported by OECD (2010), which found that if all 30 members of the OECD raised their average scores on PISA by 25 points in the next 20 years, they would gain nearly $115 trillion in GDP, during the lifetime of the post-2010 generation. Although this correlation does not necessarily imply causation, the practical implication for the US is that this modest improvement in PISA could result in a $41 trillion gain in the nation's GDP over the next 80 years. Hence, educators and policy makers should look into these international assessment data and contemplate what the US can learn from the top performers.

1.4 Responses to International Assessments

1.4.1 Hong Kong's Responses

It is not the authors' intent to downplay the merits of the US education system by introducing an overly rosy picture of some top-performing Asian countries. To offer a balanced view, the authors will discuss how some people in Hong Kong responded to international assessment results quite differently. Despite Hong Kong's outstanding performance and high ranking in international assessments, *South China Morning Post* focused on its shortcomings:

> But compared with their counterparts in Singapore – second in maths, and third in science and reading – Hong Kong pupils improved less, gaining only 11 points more in reading and six in maths and in science from the 2009 assessment. The Singaporeans gained 16 points in reading, 11 in maths and nine in science. Shanghai students also showed more improvement, gaining 14 points in reading and 13 in maths compared with 2009. The assessments also showed that just 12 percent of pupils in Hong Kong were top achievers in maths, well below the 30.8 percent for Shanghai and 19 percent in Singapore (Tam & Zhao, 2013, p. 3).

This negative tone is not surprising to the first author of this book, whose hometown is Hong Kong. There is a widespread joke about how American and Chinese parents respond to the school reports of their children differently. The following scenarios are dramatized and caricatured, but they nevertheless retain some truth. When a US student earns a B on a test, his father embraces him and says, "I am so proud of you! As long as you did your best, you ARE the best!" When a Chinese student receives an A-, his "tiger mom" yells at him, "Not an A or A + ? I'm disappointed at you! You should study harder!" There is even a joke that according to Asian standards, any grade lower than an A is technically an F. The reaction from Hong Kong's *South China Morning Post* clearly reflects how deeply the culture of fierce competition and perfectionism prevails Hong Kong society. It seems that recent international assessments have further deepened this culture.

Another article on *South China Morning Post*, entitled "Hong Kong does well in OECD table, but not all are impressed", cited the following criticism from a British educator: Hong Kong is so obsessed with grades and test results "to the extent that students are not treated as individuals but items on a production line" (Yeung, 2013, para. 5). For this critic, Hong Kong schools are like factories and examinations are nothing more than a quality control or filtering system. "Those rejected are put on a discard pile and sold off cheaply; those who pass make it to the next phase of the production line. It's not human [or] compassionate, and it's not intelligent" (Yeung, 2013, para. 5).

Nonetheless, there was still a great deal of positive feedback on Hong Kong's accomplishments in PISA. For example, Esther Ho Sui-chu, Director of Chinese University's Center for International Student Assessment, said, "I believe it shows that the city's education reform in the past 20 years has been going in the right direction and that we're not just teaching kids rote memorization" (cited in Ng, 2017, para. 21). Professor Bob Adamson, Director of Education University's Center for Lifelong Learning Research and Development, agreed with Ho that the PISA results reflected fruitful progress, but he said, "There are still areas, especially in soft skills, where there could be great improvement. Employers complain that our school graduates don't have enough creativity, entrepreneurial skills and communication skills, and are not internationalized enough. There's still a lot of work to improve our education system" (cited in Ng, 2017, para. 24).

1.4.2 US Responses

Some American researchers have rejected the notion that US students lag behind their international counterparts. For example, Berliner (2011, 2015) questioned the validity of PISA test and argued that after removing the PISA test scores of the US participants of the lowest socioeconomic status, the US students ranked the top in the world. In a similar vein, Carnoy and Rothstein (2013) argued that in PISA, disadvantaged students are oversampled from the US, and thus this type of international comparison might be unfair. Ravitch (2011) also denied the validity of those international tests, asserting, "The Soviet Union launched its Sputnik satellite in 1957. We did not respond by raising our test scores on international assessments… something is wrong with those international assessments, if our allegedly terrible public schools continue to produce the greatest workers, thinkers, leaders, and innovators that created the greatest economy in the world" (pp. 7–8).

Similarly, Wildavsky (2011) questioned whether those international assessment scores are valid indicators of US student performance. He wrote:

> But American students' performance is only cause for outright panic if you buy into the assumption that scholastic achievement is a zero-sum competition between nations, an intellectual arms race in which other countries' gain is necessarily the United States' loss. American competitive instincts notwithstanding, there is no reason for the United States to judge itself so harshly based purely on its position in the global pecking order. So long as American schoolchildren are not moving backward in absolute terms, America's relative place in global testing tables is less important than whether the country is improving teaching and learning enough to build the human capital it needs (p. 4).

More bluntly, Bracey (2009) asserted, "Using test scores to compare education systems is a mistake… using average scores to compare countries is a worse mistake" (p. 450).

Like the preceding authors, Salzman and Lowell (2008) argued that average test scores are mostly irrelevant as a measure of economic potential. Rather, high-performing students are more crucial for producing leading-edge technology. They pointed out that in PISA 2006, the US had a higher percentage of top-performing students than 5 of the 14 top-ranked countries with the highest average scores. Thus, Salzman and Lowell (2008) contended that it is unnecessary for the US to model its education after those nations that scored well in PISA. Zakaria (2015) admitted the performance gap, but he argued that although US students have not been doing well on international math and science tests since the 1960s, the US continues to dominate these fields. Thus, many authors have criticized undue attention to the average measurements in educational assessment.

To clarify, the authors believe that neither dismissing the results of these international assessments nor fostering an atmosphere of crisis with these figures is an appropriate response. To be more specific, the atmosphere of crisis in the US education system had emerged even before the rise of international assessment reports. Rothstein (1998) found that American education has been harshly criticized for over a century. For example, in 1902 the *New York Sun* asserted that the US school system did not facilitate an effective learning environment for children. In 1909 Ellwood Cubberly, the Dean of Stanford's School of Education, expressed his concern that other nations were ahead of the US in education. In 1961, a report released by the Council for Basic Education stated that about 30% ninth-graders could not read above a second- or third-grade level, thus affecting the quality of the US workforce. Rothstein argued that improvements in American education should be based on an accurate assessment of strengths and weaknesses, rather than on exaggerated criticism and blame.

1.5 Foreign Talent in the US

1.5.1 Foreign Students and the High-Tech Workforce

The authors agree with Rothstein that it is not helpful to engage in blame-and-shame tactics. We also agree with Zakaria, Salzman, and Lowell that the US will continue to take the lead in many scientific and technological fields, and we believe that scientific breakthroughs and technological innovations will likely be accomplished by a small number of elites, not necessarily by the mass of average individuals. Nonetheless, it is crucial to point out that a large portion of American elite scientists and engineers are partly educated in foreign countries. Between 1950 and 2005, 27 of the 87 American Nobel Prize winners were born outside the US (Vilcek & Cronstein, 2006). In 2003, of the 21.6 million scientists and engineers in the US, 16% (3,352,000) were immigrants (Kannankutty & Burrelli, 2007). In the same year, foreign-born doctorate holders represented approximately 50% of the workforce in engineering and computer science, as well as 37% and 43% of the workers in the physical sciences and mathematics, respectively (National Science Board, 2010). Additionally, 75% of US master's graduates in petroleum engineering are foreign students (Salzman & Lowell, 2008). In recent years approximately 71% of workforce in the Silicon Valley and 50% in the San Francisco–Oakland–Hayward region are foreign-born (Baron, 2018).

Furthermore, Hanushek (2005) pointed out that among US workers educated abroad, those from nations with higher math and science test scores consistently outperformed their counterparts who were educated in the US and other places with lower math and science scores. In a similar vein, Christensen (2008), who frequently visited Silicon Valley, asserted that the US has kept its technological edge in the world, not because its schools are sending the best-equipped technologists to US universities, but because the US has been a central magnet attracting the best talent from all corners of the world. According to the National Foundation for American Policy, international students play a crucial role in helping the US to thrive in science. Many US graduate programs have between 40% and 71% of international students (see Table 1.2). (Anderson, 2013). These international students have filled in the vacancy left by American students. In a 2011 survey conducted jointly by the University of the Sciences in Philadelphia and Harris Interactive, using a representative sample of high school students, 49% of participants said they were "definitely not" or "probably not" planning on pursuing a career in science or healthcare (Burnsed, 2011). According to a survey administered by Pew Research Center (Kennedy, Hefferon, & Funk, 2018), when the US respondents were asked why many American students do not pursue a degree in science, technology, engineering or math (STEM), 52% of them attributed the main reason to the perception that STEM subjects are too difficult.

Table 1.2 Full-time students and percent of international students in US graduate programs by field in 2011

Field	Percent of international students	Number of full-time international graduate students	Number of full-time US graduate students
Electrical Engineering	71.4	21,933	8,802
Computer Science	64.5	19,605	10,802
Industrial Engineering	60.6	4,998	3,253
Economics	55.3	7,823	6,335
Materials Engineering	53.8	3,163	2,714
Chemical Engineering	52.8	4,036	3,603
Mechanical Engineering	50.7	8,150	7,934
Mathematics and Statistics	45.6	8,354	9,949
Civil Engineering	45.6	6,554	7,809
Physics	43.6	5,844	7,569
Other Engineering	42.3	7,682	10,499
Chemistry	40.2	8,200	12,203

Source *National Science Foundation, Survey of Graduate Students and Post-doctorates in Science and Engineering*. US students include lawful permanent residents

1.5.2 STEM Exodus Among American Students

Indeed, US higher education has been suffering from low enrollment and high attrition rates in the STEM fields. According to a report from the US Department of Education's National Center for Education Statistics (Chen & Soldner, 2013), about half of the students who began on a path toward earning a bachelor's degree in science, technology, engineering, and math vacate the field before the completion of their degree. To begin with, less than 30% of students had declared a STEM major. Nearly 50% of those STEM entrants quit their majors during the course of their undergraduate years. As for those who pursued associate degrees in STEM, the statistics are even more alarming: only 20% dive into the STEM field, of whom roughly 70% retract their decision and leave the field. To be more specific, computer and information science is the major with the highest attrition rate among students who pursuing bachelor's degrees, and mathematics is the one with the highest rate among associate degree candidates. One final depressing statistic shared is that about half of those who left the STEM major ended up dropping out of their degree program, or higher education altogether. Further, the other half who had continued with their higher education in a non-STEM field showed relatively poor performances compared to their peers in the same non-STEM program.

In short, a majority of the small contingent of students who entered schools with visions of studying STEM ended up burning out and either transferring or dropping out altogether. This high attrition rate in STEM fields signals another dismal message about the current state of US education. Not only are American students lagging behind in math and science scores compared with Eastern countries, but American young adults are also exiting from the STEM fields, which naturally leads to a lack of quality influx into STEM careers. Concerned experts have sought to find the reason behind this trend. The report from the US Department of Education's National Center for Education Statistics (Chen & Soldner, 2013) conducted a series of multinomial probit (MNP) analyses to find related factors to such attrition trend. Regarding the students who left a STEM major, the difficulty level of STEM courses taken in the first year and the success level in these STEM courses showed a stronger relationship to such outcomes in comparison to many other factors, including demographics. More specifically, enrolling in easy STEM courses during the first year and then performing poorly in those STEM-related or non-STEM classes was a strong factor leading to an increased probability of quitting the STEM major. Among students who dropped out of school after departing from their STEM program, overall performance was the greatest predictor among many: poor performance as represented by lower cumulative GPA and multiple accounts of withdrawn/failed STEM courses were associated with a greater probability of dropping out.

These may explain the trend, at least partially. Nevertheless, many believe that avoiding challenging STEM courses and poor performances in STEM courses at college are not the only critical factors. The goal should not be confined to finding the reason behind it but also going back to the reason before it. One line of research is now digging up the root of the problem: what happened to students prior to their college experience regarding the math and science education they had received? Many who have explored possible reasons tend to point their fingers at K-12 math and science education (Collins & Halverson, 2018; Klein, 2003; Royster, Gross, & Hochbein, 2015; Soulé & Warrick, 2015; Venezia & Jaeger, 2013). In the 1980s, the nation suffered declining rates of college enrollment, as well as declining levels of performance in college math and science courses. The first plan of action executed by government offices, policy makers, and experts was to spawn a variety of reports and commissions to evaluate K-12 education with the purpose of reforming the preexisting educational system (Klein, 2003). The same approach is applicable to the current state of the matter with high attrition rate in STEM. The main question becomes how well primary and secondary education is preparing students for university. Statistics from the Nation's Report Card, a dissemination of several findings of the National Assessment of Educational Progress (NAEP), which has been monitoring student performance at both the national and state levels since 1969, shed light concerning the performance of young school children. Only 40% of fourth-graders and 34% of eighth-graders were found performing at or above the proficient level in mathematics. Among twelfth-graders, only 23% performed at or above the competent level in mathematics (Grigg, Donahue, & Dion, 2007). Reports revealed by the Scholastic Assessment Test (SAT) and the American College Test (ACT) indicated that less than half of high school students, usually juniors and seniors, met the College Readiness Benchmarks. Regarding the SAT, only 43% of all the class of 2012 who took the test met the SAT College & Career Readiness Benchmarks (College Board, 2012). In 2017, only 27% of all ACT-tested high school graduates were estimated as being ready for college coursework in three or four subject areas: only 41% in mathematics and 37% in science met the benchmark (ACT, Inc., 2017). Taken together, these reports offered a worrisome picture regarding the math and science education that US youth are receiving. If children are not getting the right quality of math and science education, these deficiencies would affect them when they leave secondary schools for college campuses (Bok, 2006). If students are not ready for college-level math and science, it entails a deeper and more nuanced understanding of advanced and complex problems.

On the one hand, we should celebrate the fact that the US holds a powerful international attraction. On the other hand, given the fact that foreign students tend to take STEM majors more frequently than their US peers, US schools and broader culture should face the question of why US-born students are relatively not inspired to pursue advanced math and science. Both the National Academy of Sciences (2010) and the National Science Board (2010) have warned that the US cannot reliably expect that the best from foreign nations will continue to come to the US and join the nation's STEM workforce. The real issue of the US lagging behind in science test scores may be able to stay hidden for a short time as long as immigrants continue to fill in research positions. However, Japanese-American scientist Michio Kaku has argued that without the H1-B visa program that allows foreign programmers, scientists, and engineers to work in the US, the American science establishment could collapse (Kaku, 2011). Currently, the US and China are in the race to claim the throne of the global leader in AI and machine learning. 62% of 12,521 engineering graduates who specialized in machine learning from four major Chinese universities (namely, Tsinghua University, Peking University, Shanghai Jiao Tong University, and the University of Science and Technology of China) have left China for the US, which in turn has allowed the US to retain an advantage over China in terms of the number of talented scientists, programmers, and engineers (Rayome, 2018). However, recently the relationship between China and the US has deteriorated due to the trade war and espionage scandals, which may project a foreseeable future of decreasing numbers of Chinese engineering students and graduates coming to the US.

1.5.3 International Math Olympiad

Another counterargument to the performance gap in math is based on the fact that the US team won 4th place in 2017 International Math Olympiad (IMO), 2nd place in 2015 and 1st place in 2016. In IMO, the brightest mathematics students from more than 100 countries compete by solving extremely difficult math problems. If US math education is as poor as the critics have alleged, how could the US team outperform other teams in the recent IMO? *Washington Post* commentator Strauss (2016) wrote, "Americans are generally lousy at math, right? At least that's what we hear every time there is an international test and the United States doesn't come in close to the top. But consider this: The US team of high school students just won the International Math Olympiad—said to be the most prestigious competition for high school math problem-solving." However, before 2015, the last year the US team won the IMO medal was 1995. For 20 years, the US team suffered a drought. Then, a Singaporean trainer took on the role of coaching. In fact, the US team and its coach consist almost entirely of Asian Americans. The 2015 US team was composed of Allen Liu, Yang Liu, Sammy Luo, Mark Sellke, and James Tao. The members of the 2016 US team included Ankan Bhattacharya, Michael Kural, Allen Liu, Junyao Peng, Ashwin Sah, and Yuan Yao. And the members of the 2017 team were Ankan Bhattacharya, Zachary Chroman, Andrew Gu, Vincent Huang, James Lin, and Junyao Peng (Mathematical Association of America, 2015, 2016, 2017). Indeed, in 2018 Asian Americans comprised 61–73% of the students at three elite New York schools that require high scores in the Specialized High School Admissions Test (Fuchs, 2018).

The excellent academic performances of Asian Americans are no longer a surprise to the US public. Asian Americans, including first-generation immigrants and their children, make up only 6% of the US population, yet the admission and graduation rates among Asian students in prestigious universities are exceptionally high in comparison to other ethnic groups in the US (Lee & Zhou, 2015, 2017). In Los Angeles, more than half of Asians acquired at least a college degree, in comparison to 48%

of whites, 26% of blacks, and 12% of Latinos. In prestigious public universities in California, including the University of California, Berkeley, and the University of California, Los Angeles, Asian Americans constitute more than 40% of the student body (Lee, 2018).

1.5.4 Asian Americans

Asian Americans comprise more than one-fifth of the freshmen in Ivy League universities. This pattern of success persists even after their graduation. The employment rate in Science Technology Engineering and Mathematics (STEM) field among Asians in the United States has continuously increased (Department for Professional Employees, 2013). In fact, Asians make up the largest racial minority group in the majority of STEM occupations. Between 2003 and 2013, among all STEM occupations, Asians' employment rate in computer programming jobs have increased the most in comparison to others, with an increase of 5.8%. The data also revealed that the majority of these Asian workers in STEM fields are foreigners without permanent residency or US citizenship.

Although not all Asian ethnic groups may achieve academic triumph in the US, the statistics clearly show that certain Asian groups, such as Chinese, Indians, and Koreans, are more likely to attain a successful outcome. One notable characteristic about the rate of academic success among these Asian groups is that their parents' educational background was not a contributing factor. Asian American children of working-class parents with less than a high school degree also showed a similar rate of success as their middle or high-class counterparts (Lee & Zhou, 2015, 2017). Such extraordinary academic accomplishment among Asian Americans has led many scholars and analysts to search for the reason behind this success.

The report *Asian American Achievement Paradox* (2015) asserted that this remarkable academic success among Asians may be traced back to hyper-selectivity among Asian immigrants. Asian immigrants bring their *success frame* (p. 18), a mindset and plan of action for academic and professional success, from their mother countries. Asian immigrants and their children execute this success frame, faithfully following the road of academic excellence, admission to top-tier schools, and landing in STEM-related professions. Success frame also includes building a strong academic support system such as tutoring, private academies, and high-quality prep courses within tight-knit communities. They reassemble these institutions and make these services available to almost every child. Even the children of working-class parents who own a small laundry shop have access to these academic resources. Most parents are active members of their Asian immigrant communities with a concrete success frame and hold to the importance of education and its impact on their children's futures. The success frame is circulated and reinforced among Asian immigrants and their children, which then inevitably contributes to their success in academics.

1.6 What Can Be Done? Malleable Factors and Actionable Items

Although some educators have acknowledged the performance gap, they have been skeptical or critical of the proposal of emulating Asian practices. For example, Menon (2000) argued that some schools in Asia assign students into different tracks when they are still very young. Those who have earned high academic scores are placed on the path toward university education, while those who have not done as well are put into vocational tracks. Menon's challenge implies that the Asian approach is incompatible with the American philosophy that everyone is entitled to have freedom of choice and an equal opportunity to attain his or her goals. Furthermore, some high school courses in Singapore are equivalent to college-level courses in America. In order to equip their children with test-taking skills for the difficult curriculum, many East Asian parents hire tutors for their children and demand that they devote long hours to studying. Menon questioned whether it would be beneficial for the US to insist on making things very challenging for all students. Likewise, Atkinson and Mayo (2010) are opposed to the ideas of imposing more rigorous STEM standards and more assessments to all students. Nevertheless, it is the conviction of the authors that US educators can still learn from those Asian countries, including Hong Kong, in spite of cultural differences. After all, the Hong Kong education model heavily borrowed ideas from the West, including the UK and the US. More details will be discussed in Chap. 6, "Cultural Difference and Change Model".

Numerous studies have been conducted with the goal of learning from the best practices of Asian educational systems. Although several theories exist to explain why Asian students excel in science, some of those theories do not address malleable factors, and as a result, they do not necessarily lead to actionable items and practical applications. For example, some authors attribute Asians' academic success to parents' socioeconomic status (SES), parenting styles, and the prevailing cultural value on education (Ho, 2009, 2010a, 2010b; Otto, 2016; Tan, 2012). Abbound and Kim (2005) asserted that, while US children are distracted from academic work by many extracurricular activities and household chores, Asian students are encouraged by their parents to concentrate their efforts solely on their schoolwork. The role of Asian children in the family is simply to respect the elders, to obey their parents, and to study hard in order to secure a bright future. In a similar vein, using data from Hong Kong, Japan, Korea, Singapore, and Thailand from TIMSS 1995, Wößmann (2005) found that family background played an important role in affecting student test performance. The importance of family background was further supported by Hojo and Oshio (2010) in their analysis of the data of Hong Kong, Japan, Sweden, Singapore, South Korea, and Taiwan from TIMSS 2007. Additionally, in her controversial popular book, Chua (2011) asserted that Chinese parents could raise academically successful children due to the "Tiger Mom" approach. For example, she never allowed her children to attend a sleepover, to have a playmate, to be in a school play, to watch TV or play video games, to choose extracurricular activities on their own, to get any grade less than an A, or to be less than number one in every subject (except gym and drama). Some researchers went even further to explain these East–West differences from a macro-cultural perspective. For instance, based on the data obtained from the 2000 and 2003 Programme for International Student Assessment (PISA), Ho (2009) found that there are some primary cognitive habits that are shared by the Confucian Heritage Culture shared by East Asian societies.

Some authors (e.g., Levitt & Dubner, 2014) went even further to argue that school-based reforms, such as improving teachers' skills, are not effective ways to raise academic standards because teachers' skills have less influence on a student's performance than parenting styles do. Specifically, children stay at school for only seven hours a day, 180 days a year, or about 22% of their waking hours. On the contrary, parenting styles have much more impact on the development of children's mentality and habits.

Although these explanations have theoretical significance, it is difficult for policy makers to design a plan to promote the Asian parenting style. By the same token, it is also impractical to transform a cultural system (e.g., infusing Confucian cultural elements into the US in a short amount of time). Strictly speaking, those arguments fail to provide *actionable items*. In his comments on the American parenting styles, Levitt & Dubner (2014) writes, "In our society, if someone wants to be a hairstylist or a kickboxer or a hunting guide—or a schoolteacher—he or she must be trained and licensed by a state agency. No such requirement is necessary for parenthood. Anyone with a set of reproductive organs is free to create a child" (p. 30). If any government agency were to administer a parenting licensure exam or attempt to persuade people to adopt features of an Asian parenting style, such as urging parents not to allow their children to have a playmate or ban all computer games, we can imagine an immediate and widespread protest.

Thus, it is more fruitful to address only malleable factors at this moment. According to Rudy and Sussman (2013), a malleable factor is a changeable factor under the control of the educational system, including student characteristics (e.g., student behavior, student skills, and student knowledge), teacher characteristics (e.g., teaching style and credentials), school characteristics (e.g., class size and school climate), and actual education interventions (e.g., curriculum, instructional approaches, and education programs).

The authors strongly believe that the pedagogical approach of teachers still plays a significant role in education reform because it is a malleable factor. For example, in a study of Hong Kong eighth-grade students based on TIMSS 1999 data, House (2000) found that having the teacher ask the students what they knew about the new topic, and explain the rules and definitions afterward, was related to higher science test scores. Likewise, students who reported that they spent more time conducting an experiment, engaging in practical investigation in class, and watching a demonstration of an experiment tended to result in higher science test scores. In addition, students who worked together in pairs or small teams in their science lessons were more likely to earn higher science test scores. The same conclusion was reached in another study using Japanese fourth-grade student data obtained from TIMSS 1995 (House, 2006). Schools can offer workshops and seminars to teachers regarding pedagogical methods. In this book, the authors focus on why American teachers should reexamine their existing approach, and also consider what they can learn from Asian educational systems, especially that of Hong Kong.

1.7 Methodology

As mentioned above, some researchers are critical of using test scores to drive education policy. As a remedy to the limitations of quantitative studies, the authors have adopted a triangulation mixed method-design (Creswell, 2010), in which both quantitative and qualitative data are analyzed concurrently so that a comprehensive and coherent picture of the phenomenon can be drawn. Our quantitative data are sourced from international assessments, including PISA, PIAAC, and TIMSS. Needless to say, the sample size of those assessments is considerably large, and thus conventional statistical procedures are inappropriate. To rectify this situation, data mining, data visualization, and other modern analytical methods were employed. For example, in the analysis of PISA data regarding the relationship between academic performance and self-image, data visualization and nonlinear modeling (rather than OLS regression) were utilized for pattern recognition. In analyzing TIMSS data to scrutinize the association between instrumental motivation and science learning, jackknife regression was employed due to the nested data structure of TIMSS.

Our qualitative data result from either online or face-to-face interviews with 35 math and science teachers and 54 professors and lecturers of math and science education in Hong Kong. For teacher interviews, the focal points centered on pedagogical techniques, especially lessons, exercises, and activities specific to problem-solving and creativity. Regarding the interviews of professors and lecturers, the emphasis was placed on their research into math and science education, and also their perceptions of current science and math education in Hong Kong. There was no age or nationality restriction for teacher and professor/lecturer samples. As long as they currently teach math/science at high school or are involved in math/science education (e.g., conducting research, providing training), their feedback was considered for the analysis. The research team followed the sampling scheme of grounded theory. Hence, there was no predetermined sample size. Rather, the team kept collecting data until the point of saturation, where no additional insight could be uncovered by additional data (Charmaz, 2014). In the process of collecting samples, we also utilized snowball sampling: recruiting more subjects from their acquaintances (Goodman, 1961). At first glance, this sampling method may seem arbitrary. Nonetheless, the authors considered the results of this exploratory approach to be serendipitous. The authors are well aware that this sample might not be representative of the entire Hong Kong population. Nonetheless, the purpose of qualitative data collection is to provide a nuanced and detailed view of a limited group. Second, whether the pedagogy mentioned in this book could represent the whole teacher population in Hong Kong was not the main objective of the authors. Our goal was and is to learn from the best, not from the

Table 1.3 The affiliation of the faculty that participated in the project

University	Count	Percentage
Chinese University of Hong Kong	6.00	0.11
City University of Hong Kong	5.00	0.09
Education University of Hong Kong	12.00	0.22
HK Academy for Gifted Education	1.00	0.02
Hong Kong Baptist University	5.00	0.09
Hong Kong Institute of Vocational Education	1.00	0.02
Hong Kong Polytechnic University	6.00	0.11
Hong Kong University of Science and Technology	6.00	0.11
Lingnan University	2.00	0.04
University of Hong Kong	8.00	0.15
Others	2.00	0.04

average. Due to the legal restrictions, the authors were not allowed to disclose any identifiable demographic profiles of interviewed Hong Kong teachers, such as their credentials, or any identifiable school characteristics, such as ranking (e.g., Band 1, Band 2, etc.). Nonetheless, the demographics of Hong Kong professors are reported as follows: the mean age of the faculty was 48.5, and the average years of teaching experience were 13. Thirteen of them were females (24%) and 41 were males (76%). Every university or postsecondary institution in Hong Kong has a voice in the sample (see Table 1.3).

Although it is a common practice for grounded theorists to extract common themes and categories from interview data and open-ended responses, the research team did not consider this approach to be a good fit for this study. First, grounded theorists must put aside all preconceptions so that there is room for fresh and new categories to emerge from the data (Charmaz, 2014). However, the objective of this study was not to discover new concepts or to generate a novel hypothesis. Rather, we have already acknowledged the overarching themes by reviewing the policy statements announced by the Hong Kong Education Bureau and Hong Kong Examinations and Assessment Authority (see Chap. 2, e.g., learning to learn, problem-solving, critical thinking, creativity, etc.). In addition, the authors could access numerous books focusing on the theoretical aspects of math and science education. Thus, the authors turned to concrete examples, such as specific lessons, exercises, and activities that facilitate lifelong learning, problem-solving, and creativity. Based on our initial reading of the data, we have decided that counting the recurrence frequency of abstract concepts was not a fruitful way to approach the problem. Overall, most teachers and professors agreed that promoting creativity and problem-solving skills is important. In the end, the team decided to present those lessons, exercises, and activities by grouping them into preset categories (e.g., problem-solving, creativity, etc.). By analyzing these unstructured materials, the authors hope to provide US educators with concrete examples of Hong Kong pedagogical approaches.

1.8 Organization of the Book

The subsequent chapters will illustrate how we can learn from Hong Kong by providing the readers with concrete examples. Chapter 2 introduces the overarching learning goals (e.g., learning to learn) and the assessment framework developed by the Hong Kong Education Bureau. It is important to point out that these guidelines and objectives are not empty slogans. Rather, many Hong Kong educators take them seriously by developing innovative pedagogical methods and implementing them in their classrooms as well. Chapter 3 discusses the relationship between creativity, personality traits, and basic knowledge. Tangible examples for promoting creativity are provided. Contrary to a popular belief, the current Hong Kong learning approach no longer emphasizes rote learning and exam outcomes. Students who merely follow the rules would be incapable of solving real-life problems, which are usually ill-defined and ambiguous. Chapter 4 focuses on how Hong Kong teachers enhance the problem-solving skills of learners by making instructions vague, incomplete, and difficult. In addition, problem-solving is a tedious process of trial and error; hence it requires one's humility and an appropriate assessment of one's self-image to accept failures. Thus, Chap. 5 discusses how productive failure can reshape students' self-efficacy. However, neither meaningful nor fruitful learning can be achieved if students themselves cannot see the true value of lifelong learning, creativity, and problem-solving. Therefore, Chap. 6 shifts gear to examine the motivational aspects of Hong Kong students. The last chapter is a summary about how we can adapt the preceding approaches into another cultural context, more specifically, from Hong Kong to the US.

References

Abbound, S. K., & Kim, J. Y. (2005). *Top of the class: How Asian parents raise high achievers—And how you can too*. New York, NY: Berkley Books.

American College Testing [ACT], Inc. (2017). *The condition of college & career readiness 2017*. Retrieved from http://www.act.org/content/dam/act/unsecured/documents/cccr2017/CCCR_National_2017.pdf.

Anderson, S. (2013). The importance of international students to America. NEAP policy report brief. *National Foundation for American Policy*. Retrieved from http://www.nfap.com/pdf/New NFAP Policy Brief The Importance of International Students to America, July 2013.pdf.

Ang, R. P., & Huan, V. S. (2006). Relationship between academic stress and suicidal ideation: Testing for depression as a mediator using multiple regression. *Child Psychiatry and Human Development, 37*(2), 133–143. https://doi.org/10.1007/s10578-006-0023-8.

Atkinson, R. D., & Mayo, M. (2010). *Refueling the US innovation economy: Fresh approaches to science, technology, engineering, and technology (STEM) education*. Washington, DC: Information Technology and Innovation Foundation. Retrieved from https://itif.org/files/2010-refueling-innovation-economy.pdf.

Baron, E. (2018, January 17). H-1B: Nearly three-quarters of Silicon Valley tech workers foreign born. *Silicon Beat*. Retrieved from http://www.siliconbeat.com/2018/01/17/h-1b-nearly-three-quarters-silicon-valley-tech-workers-foreign-born-report-says/.

Berliner, D. C. (2011). The context for interpreting PISA results in the USA. Negativism, chauvinism, misunderstanding, and the potential to distort the educational systems of nations. In M. Pereira, H-G. Kotthoff, & R. Cowen (Eds.), *PISA under examination: Changing knowledge, changing tests, and changing schools* (pp. 77–96). Rotterdam: Sense Publishers.

Berliner, D. C. (2015). The many facets of PISA. *Teachers College Record, 117*(1), 1–20.

Bjorklund, B. R. (2015). *The journey of adulthood* (8th ed.). Boston, MA: Pearson.

Bok, D. (2006). *Our underachieving colleges: A Candid look at how much students learn and why they should be learning more*. Princeton, NJ: Princeton University Press.

Bracey, G. (2009). PISA: Not leaning hard on US economy. *Phi Delta Kappan, 90,* 450–451.

Burnsed, B. (2011, May). Combating students' disinterest in the sciences: Students show diminishing desire to pursue a career in the sciences despite rosy job prospects. *US News and World Report*. Retrieved from https://www.usnews.com/education/best-colleges/articles/2011/05/23/combating-students-disinterest-in-the-sciences.

Calmes, J. (2010, December 6). Obama calls for new Sputnik moment. *New York Times*. Retrieved from http://thecaucus.blogs.nytimes.com/2010/12/06/obama-calls-for-new-sputnik-moment/.

Carnoy, C., & Rothstein, R. (2013). What do international tests really show about US. student performance? *Economic Policy Institute*. Retrieved from http://www.epi.org/publication/us-student-performance-testing/.

Charmaz, K. (2014). *Constructing grounded theory*. Thousand Oaks, CA: Sage.

Chen, X. L., & Soldner, M. (2013). *STEM attrition: College students' paths into and out of STEM fields: Statistical analysis report*. Washington, DC: National Center for Education Statistics.

Cheo, R., & Quah, E. (2005). Mothers, maids, and tutors: An empirical evaluation of their effect on children's academic grades in Singapore. *Education Economics, 13*(3), 269–285.

Cheong, C. Y. (2009). Hong Kong educational reforms in the last decade: Reform syndrome and new developments. *International Journal of Educational Management, 23*(1), 65–86.

Cheow, S.A. (2019, April 12). *More teens in Singapore seeking help at IMH for school stress*. Retrieved from https://www.straitstimes.com/singapore/education/more-teens-in-singapore-seeking-help-for-school-stress-at-imh.

Christensen, C. M. (2008). *Disrupting class: How disruptive innovation will change the way the world learns*. New York: McGraw-Hill.

Chua, A. (2011). *Battle hymn of the Tiger Mother*. New York, NY: Penguin Press.

College Board. (2012). *SAT Report: Only 43 percent of 2012 college-bound seniors are college ready*. New York, NY: College Board. Retrieved from http://media.collegeboard.com/digitalServices/pdf/research/sat-report-2012-press-release.pdf.

Collins, A., & Halverson, R. (2018). *Rethinking education in the age of technology: The digital revolution and schooling in America*. New York, NY: Teachers College Press.

Columbus, L. (2017, May 13). IBM predicts demand for data scientists will soar 28% by 2020. *Forbes*. Retrieved from https://www.forbes.com/sites/louiscolumbus/2017/05/13/ibm-predicts-demand-for-data-scientists-will-soar-28-by-2020/#444c32257e3b.

Creswell, J. (2010). *Designing and conducting mixed methods research* (2nd ed.). Thousand Oaks, CA: Sage.

Csapó, B., & Funke, J. (Eds.). (2017). *The nature of problem solving: Using research to inspire 21st century learning, educational research and innovation*. Paris: OECD Publishing. https://doi.org/10.1787/9789264273955-en.

Deisenroth, M. P., Faisal, A. A., & Ong, C. S. (2019). *Mathematics for machine learning*. Cambridge, UK: Cambridge University Press.

Department for Professional Employees. (2013). *The Asian and Hawaiian/Pacific Islander STEM workforce*. Washington, DC: Author. Retrieved from http://dpeaflcio.org/programs-publications/issue-fact-sheets/the-asian-and-hawaiianpacific-islander-stem-workforce/.

Emeagwali, N. S. (2010). National science board says US lead in STEM slipping. *Techniques: Connecting Education & Careers, 85*(3), 10–11.

Fisher, S. (1994). *Stress in academic life: The mental assembly line*. London, UK: Open University Press.

Forlin, C. (2010). Developing and implementing quality inclusive education in Hong Kong: Implications for teacher education. *Journal of Research in Special Educational Needs, 10,* 177–184.

Fuchs, C. (2018, June 18). Plan to diversify New York's top high schools divides Asian-American groups. *NBC News*. Retrieved from https://www.nbcnews.com/news/asian-america/plan-diversify-new-york-s-top-high-schools-divides-asian-n884316.

Goodman, L. A. (1961). Snowball sampling. *Annals of Mathematical Statistics, 32*(1), 148–170. https://doi.org/10.1214/aoms/1177705148.

Grigg, W., Donahue, P., & Dion, G. (2007). The Nation's report card: 12th-grade reading and mathematics 2005 (NCES 2007-468). National Center for Education Statistics. Washington, DC: US. Government Printing Office.

Grobler, G. C. (2018). STEM in a day. *Hong Kong Science Teachers' Journal, 33,* 21–25.

Gustafsson, J. F. (2016). Lasting effects of quality of schooling: Evidence from PISA and PIAAC. *Intelligence, 57,* 66–72.

Hanushek, E. A. (2005, June). Why quality matters in education. *Finance and development, 42*(2), 15–19.

Hanushek, E. A., & Woessmann, L. (2008). The role of cognitive skills in economic development. *Journal of Economic Literature, 46,* 607–668.

Hui, A. N. N., & Lau, S. (2010). Formulation of policy and strategy in developing creativity education in four Asian Chinese societies: A policy analysis. *Journal of Creativity Education, 44,* 215–235.

Ho, E. S. (2009). Characteristics of East Asian learners: What we learned from PISA. *Educational Research Journal, 24,* 327–348.

Ho, E. S. (2010a). Family influences on science learning among Hong Kong adolescents: What we learned from PISA. *International Journal of Science and Mathematics Education, 8,* 409–428.

Ho, E. S. (2010b). Assessing the quality and equality of Hong Kong basic education results: From PISA 2000 + to PISA 2006. *Frontiers of Education in China, 5,* 238–257.

Hojo, M., & Oshio, T. (2010). What factors determine student performance in East Asia? New evidence from TIMSS 2007. *IDEAS, Department of Economics, University of Connecticut.* Retrieved from http://ideas.repec.org/p/hit/piecis/494.html.

House, J. D. (2000). Relationship between instructional activities and science achievement of adolescent students in Hong Kong: Findings from the Third International Mathematics and Science Study (TIMSS). *International Journal of Instructional Media, 27,* 275–288.

House, J. D. (2006). Mathematics beliefs, instructional strategies, and algebra achievement of adolescent students in Japan: Results from the TIMSS 1999 assessment. *International Journal of Instructional Media, 33,* 443–462.

Huan, V. S., See, Y. L., Ang, R. P., & Har, C. W. (2008). The impact of adolescent concerns on their academic stress. *Educational Review, 60*(2), 169–178.

Hunt, E., & Wittmann, W. (2008). National intelligence and national prosperity. *Intelligence, 36*(1), 1–9.

Isaacson, W. (2017). *Leonardo da Vinci*. New York, NY: Simon & Schuster.

Joncas, M. (2008). TIMSS 2007 sample design. In J. F. Olson, M. O. Martin, & I. V. S. Mullis (Eds.), *TIMSS 2007 technical report* (pp. 77–92). Chestnut Hill, MA: TIMSS & PIRLS International Study Center, Boston College.

Jyoti, S., & Neetu, S. (2013). Implications of corporal punishment on primary school children. *IOSR Journal of Humanities and Social Science, 15*(6), 57–61.

Kaku, M. (2011). *America has a secret weapon* [Video file]. Retrieved from https://www.youtube.com/watch?v=NK0Y9j_CGgM.

Kannankutty, N., & Burrelli, J. (2007, June). Why did they come to the United States? A profile of immigrant scientists and engineers. *National Science Foundation Info Brief, NSF07-324.* Retrieved from http://www.nsf.gov/statistics/infbrief/nsf07324/.

Kennedy, B., Hefferon, M., & Funk, C. (2018, January). Half of Americans think young people don't pursue STEM because it is too hard. *Pew Research Center*. Retrieved from https://www.pewresearch.org/fact-tank/2018/01/17/half-of-americans-think-young-people-dont-pursue-stem-because-it-is-too-hard/.

Kepner, J., & Jananthan, H. (2018). *Mathematics of big data: Spreadsheets, databases, matrices, and graphs*. Boston, MA: MIT Press.

Klein, D. (2003). A brief history of American K-12 mathematics education in the 20th century. In J. M. Royer (Ed.), *Mathematical cognition* (pp. 175–259). Charlotte, NC: Information Age Publishing.

Kornblut, A. E., & Wilson, S. (2011, January 26). State of the Union 2011: 'Win the future,' Obama says. *Washington Post*. Retrieved from http://www.washingtonpost.com/wp-dyn/content/article/2011/01/25/AR2011012504068.html.

Kung, D. (2013). *How music and mathematics relate*. Chantilly, VA: The Teaching Company.

Law, W. W. (2004). Translating globalization and democratization into local policy: Educational reform in Hong Kong and Taiwan. *International Review of Education, 50,* 497–524.

Lee, J. (2018, January 23). It takes more than grit: Reframing Asian American academic achievement. *Social Science Research Council*. Retrieved from https://items.ssrc.org/it-takes-more-than-grit-reframing-asian-american-academic-achievement/.

Lee, J., & Zhou, M. (2015). *The Asian American achievement paradox*. New York, NY: Russell Sag Foundation.

Lee, J., & Zhou, M. (2017). Why class matters less for Asian-American academic achievement. *Journal of Ethnic and Migration Studies, 43*(14), 2316–2330.

Leshner, A. (2009, January 12). A wake-up call for science education. *Boston Globe*. Retrieved from http://www.boston.com/bostonglobe/editorial_opinion/oped/articles/2009/01/12/a_wake_up_call_for_science_education.

Levitt, S. D., & Dubner, S. J. (2014). *Think like a freak: The authors of freakonomics offer to retrain your brain*. New York, NY: William Morrow.

Linton, J. (2010). United States department of education update. *Journal of Correctional Education, 61*(1), 6–8.

Mathematical Association of America. (2015). *US team takes second at International Mathematical Olympiad*. Retrieved from https://www.maa.org/news/us-team-takes-second-at-international-mathematical-olympiad.

Mathematical Association of America. (2016). *US wins first place at International Mathematics Competition in Hong Kong*. Retrieved from https://www.maa.org/news/us-wins-first-place-at-international-mathematics-competition-in-hong-kong.

Mathematical Association of America. (2017). *US students bring home medals at International Mathematics Competition in Brazil*. Retrieved from https://www.maa.org/news/us-students-bring-home-medals-at-international-mathematics-competition-in-brazil.

Mathematical Association of America. (2019). *Symphonic equations: A mathematical exploration of music*. Retrieved from https://www.maa.org/meetings/calendar-events/maa-distinguished-lecture-series/math-and-music-enthrall-at-maa-distinguished-lecture.

McGonigal, K. (2016). *The upside of stress: Why stress is good for you, and how to get good at it*. London, UK: Penguin.

Menon, R. (2000). Should the United States emulate Singapore's education system to achieve Singapore's success in the TIMSS? *Mathematics Teaching in the Middle School, 5,* 345–347.

National Academy of Sciences. (2010). *Expanding underrepresented minority participation: America's science and technology talent at the crossroads*. Washington, DC: National Academies Press.

National Science Board. (2010). *Preparing the next generation of STEM innovators: Identifying and developing our nation's human capital*. Arlington, VA: National Science Foundation.

Ng, M. (2017, November 21). Hong Kong pupils among world's best group problem-solvers (but Singapore tops the chart): Results from Pisa assessment of how well students work together for solutions show Japan in second. *South China Morning Post*. Retrieved from https://www.scmp.com/news/hong-kong/education/article/2120897/hong-kong-pupils-among-worlds-best-group-problem-solvers.
Ngiam, X. Y., & Tung, S. S. (2016). The acceptability of caning children in Singapore: the fine line between discipline and physical maltreatment. *Journal of Developmental and Behavioral Pediatrics, 37*(2), 158–163.
Organization for Economic Cooperation and Development [OECD]. (2007). *PISA 2006: Science competencies for tomorrow's world*. Paris: Organization for Economic Cooperation and Development [OECD].
Organization for Economic Cooperation and Development [OECD]. (2010). *The high cost of low educational performance: The long run economic impact of improving PISA outcomes*. Paris: Organization for Economic Cooperation and Development [OECD]. Retrieved from http://www.oecd.org/dataoecd/11/28/44417824.pdf.
Organization for Economic Cooperation and Development [OECD]. (2011). *Lessons from PISA for the United States–Shanghai and Hong Kong: Two distinct examples of education reform in China*. Paris: Author. https://dx.doi.org/10.1787/9789264096660-5-en.
Organization for Economic Cooperation and Development [OECD]. (2012). *Literacy, numeracy and problem solving in technology-rich environments framework for the OECD survey of Adult Skill*. https://doi.org/10.1787/9789264128859-en. Retrieved from http://www.oecd.org/skills/piaac/PIAAC%20Framework%202012–%20Revised%2028oct2013_ebook.pdf.
Organization for Economic Cooperation and Development [OECD]. (2013a). *Technical report of the survey of adult skills (PIAAC)*. Retrieved from https://www.oecd.org/skills/piaac/_Technical%20Report_17OCT13.pdf.
Organization for Economic Cooperation and Development [OECD]. (2013b). Students' drive and motivation. Retrieved from http://www.oecd.org/pisa/keyfindings/PISA2012-Vol3-Chap3.pdf.
Organization for Economic Cooperation and Development [OECD]. (2016). *Skills matter: Further results from the survey of adult skills*. Retrieved from http://www.keepeek.com/Digital-Asset-Management/oecd/education/skills-matter_9789264258051-en#.V-ZEjpMrKL4.
Otto, W. (2016). What teachers should know about why these students perform so well: An examination of Korean-American achievement through student perspectives of East Asian parenting beliefs, styles, and practices. *International Electronic Journal of Elementary Education, 9*(1), 167–181.
Parsons, S., & Bynners, J. (2005). *Does numeracy matter more?*. London, UK: National Research and Development Centre for Adult Literacy and Numeracy.
Pearson, E., & Rao, N. (2006). Early childhood education policy reform in Hong Kong: Challenges in effecting change in practices. *Childhood Education, 82*, 363–368.
Ravitch, D. (2011, February). *Keynote address*. Speech presented at the Annual Meeting of the American Association of School Administrators, Denver, CO. Retrieved from http://www.aasa.org/uploadedFiles/NCE/The_Conference_Daily_Online/2011_Conference_Daily/2011_Conference_Daily_Day_3_February_18,_2011/DianeRavitchSpeech-NCE11-text-format.pdf.
Rayome, A. D. (2018, December 4). 62% of China's machine learning graduates leave to work in the US. *Tech Republic*. Retrieved from https://www.techrepublic.com/article/62-of-chinas-machine-learning-graduates-leave-to-work-in-the-us/.
Renzulli, K. A., Weisser, C., & Leonhardt, M. (2016, May 16). The 21 most valuable career skills now. *Money*. Retrieved from http://time.com/money/4328180/most-valuable-career-skills/.
Root-Bernstein, R., & Root-Bernstein, M. (2004). Artistic scientists and scientific artists: The link between Polymathy and creativity. In R. J. Sternberg, E. L. Grigorenko, & J. L. Singer (Eds.), *Creativity: From potential to realization* (pp. 127–151). Washington, DC, US: American Psychological Association. http://dx.doi.org/10.1037/10692-008.

Rothstein, R. (1998). *The way we were? The myths and realities of America's student achievement*. New York, NY: Century Foundation Press.

Royster, P., Gross, J., & Hochbein, C. (2015). Timing is everything: Getting students back on track to college readiness in high school. *The High School Journal, 98,* 208–225.

Rudy, A., & Sussman, A. (2013). IES grant writing workshop for exploration projects. Retrieved from https://ies.ed.gov/funding/webinars/pdf/WODexploration_FY2013.pdf.

Salzman, H., & Lowell, L. (2008). Making the grade. *Nature, 453,* 28–30.

Sing, C. C., & Khine, M. S. (2008). Assessing the epistemological and pedagogical beliefs among pre-service teachers in Singapore. In M. S. Khine (Ed.), *Knowing, knowledge and beliefs* (pp. 287–299). Singapore: Springer.

Soulé, H., & Warrick, T. (2015). Defining 21st century readiness for all students: What we know and how to get there. *Psychology of Aesthetics, Creativity, and the Arts, 9*(2), 178–186.

Strauss, V. (2016, July 18). U.S. students win prestigious International Math Olympiad for second straight year. *Washington Post*. Retrieved from https://www.washingtonpost.com/.

Tam, J., & Zhao, S. (2013, December 3). Shanghai teens still world's best at reading, maths, science in Pisa survey. *South China Morning Post*. Retrieved from https://www.scmp.com/news/china/article/1372060/shanghai-teens-still-worlds-best-reading-maths-science-pisa-survey.

Tan, C. (2011). Framing educational success: A comparative study of Shanghai and Singapore. *Education, Knowledge and Economy, 5*(3), 155–166.

Tan, C. (2012). "Our shared values" in Singapore: A confucian perspective. *Educational Theory, 62,* 449–463.

Tan, C. (2017). Private supplementary tutoring and parentocracy in Singapore. *Interchange, 48,* 315–329.

Trump, D. (2019, February 11). Executive order on maintaining American leadership in artificial intelligence. *The White House*. Retrieved from https://www.whitehouse.gov/presidential-actions/executive-order-maintaining-american-leadership-artificial-intelligence/.

U.S Department of Labor. (2017). Employment projections: Fastest growing occupations. *Bureau of Labor Statistics*. Retrieved from https://www.bls.gov/ooh/fastest-growing.htm.

U.S Department of Labor. (2019). *Labor statistics*. Retrieved from https://www.bls.gov/.

Venezia, A., & Jaeger, L. (2013). Transitions from high school to college. *The Future of Children, 23*(1), 117–136.

Vilcek, J., & Cronstein, B. N. (2006). A prize for the foreign-born. *Journal of the Federation of American Societies for experimental Biology, 20,* 1281–1283.

Wang, J. (2018). Innovation and government intervention: A comparison of Singapore and Hong Kong. *Research Policy, 47,* 399–412. https://doi.org/10.1016/j.respol.2017.12.008.

Weiss, S. F., Murray, R. E., & Cyrus, C. J. (2010). *Music education in the Middle Ages and the Renaissance*. Bloomington, IN: Indiana University Press.

West, M. R. (2012). Global lessons for improving US education. *Issues in Science & Technology, 28*(3), 37–44.

Wildavsky, B. (2011, February 21). Relax, America. Chinese math whizzes and Indian engineers aren't stealing your kids' future. *Think Again: Education*. Retrieved from https://foreignpolicy.com/2011/02/21/think-again-education/.

Wiseman, A. (2015). *How the world learns: Comparative educational systems*. Chantilly, VA: The Teaching Company.

Wößmann, L. (2005). Educational production in East Asia: The impact of family background and schooling policies on student performance. *German Economic Review, 6,* 331–353.

Wong, T. H. (2015). Social foundations of public–private partnerships in education: The historical cases of post-war Singapore and Hong Kong. *History of Education, 44,* 207–224. https://doi.org/10.1080/0046760X.2014.979256.

Yeung, L. (2013, December 9). Reflections: Hong Kong does well in OECD table, but not all are impressed. *South China Morning Post*. Retrieved from https://www.scmp.com/lifestyle/family-education/article/1374567/reflections-hong-kong-does-well-oecd-table-not-all-are.

Yu, C. H., Lee, H. S., Lara, E., & Gan, S. G. (2019). Adult learning across USA, Canada, and New Zealand: A cross-cultural study of PIAAC. *International Education Studies, 12*(5), 1–16. https://doi.org/10.5539/ies.v12n5p1. Retrieved from http://www.ccsenet.org/journal/index.php/ies/article/view/0/39300.

Zakaria, F. (2015). *In defense of a liberal education*. New York, NY: W. W. Norton.

Chapter 2
Overarching Framework: Learning to Learn

While Germany and Japan strongly responded to their unexpected low scores after their participation in PISA in 2000 and 2003, respectively, the first noticeable public discourse within the US about international assessments happened in 2010 (Martens & Niemann, 2013; Wiseman, 2015). It is encouraging that the US has now finally investigated the educational system of those nations that outperformed America (Duncan, 2011). As former U.S. Secretary of Education Duncan (2010) said, "The simple truth is that America has a great deal to learn from the educational practices of other countries" (p. 73). What can we learn from our international counterparts? In particular, being faithful to the focus of this book, what can we learn from evaluating the educational policies and practices of Hong Kong? This chapter analyzes the major components of the Hong Kong educational system, including a mindset of learning to learn, an endeavor for problem-solving, and a mastery of skills.

2.1 Hong Kong Education Framework

While presenting a proposal for educational reform to the UK government, the UK Secretary of State for Education praised the Hong Kong model as "one of the world's most successful school systems" (Gove, 2013). It is interesting for the UK Secretary of State for Education to praise a former colony of the British Empire because the Hong Kong approach, to a large extent, was modeled after the UK system. Under British colonial rule between 1842 and 1997, Hong Kong's education system was developed and fine-tuned by policy makers from the UK. In 1997, the People's Republic of China resumed sovereignty over Hong Kong, accepting the model of "one country with two systems". China did not impose its education system on Hong Kong, and thus Hong Kong managed to enjoy a high degree of autonomy in education policy. On the one hand, Hong Kong maintained the UK infrastructure in education, but on the other hand, Hong Kong radically reformed its education. The UK was a major source of expertise for Hong Kong's post-1997 education reform. The new school evaluation system borrowed from the Scottish model is one

C. H. Yu and H. S. Lee, *Creating Change to Improve Science and Mathematics Education*, https://doi.org/10.1007/978-981-15-3156-9_2

of many examples. To accommodate learner differences, Hong Kong reformers also borrowed from UK-based academics to include the linking of curriculum, assessment, school accountability, and varied pedagogical approaches. More options for vocational training, which is similar to England's Business, Technology, and Education Council qualifications, were offered to less academically oriented students (Forestier & Crossley, 2015).

Thus, it is not surprising to see that many good ideas observed in Hong Kong education can also be found in England and other Western countries, yet we can still learn from Hong Kong in its refinement and implementation of those ideas. Following this line of reasoning, the authors strongly support the strategy of multi-way borrowing, as implied by Wiseman's research (2015). After comparing various education practices around the world, Wiseman (2015) found that many best practices used by top-performing countries in PISA are also found in the US; nonetheless, we need to understand how and why those methods work so well in other cultural contexts.

Unlike the US, where education systems vary from state to state, Hong Kong education framework is centralized by the Education Bureau. In all subject matters, the Hong Kong Education Bureau (HKEDB) (2018) intends to provide quality school education for Hong Kong students, to develop their potential to the fullest, and to prepare them for life's challenges by delivering professional services, ensuring the effective use of resources, and forging partnerships to promote excellence in school education.

2.1.1 Exam-Oriented Culture and the Through Road Approach

One common reaction to learning from Hong Kong is the concern of importing an exam-oriented culture. This skeptical attitude is well justified. In the past, under a highly competitive exam-oriented education system, the entire life of students was occupied by studying. In the end, only a few could pass the public exam and enjoy a brighter future, while those who failed were deprived of further opportunities. For example, in 1975 only 2% of secondary school graduates were admitted to universities, meaning that only the top performers in the public exam could have the chance to receive higher education. Hong Kong educators were well aware of this problem. The Education Commission (1999) described the Hong Kong education system in the past as follows:

> Our education system appears to have stagnated in the industrial age. The system still caters to a selected few, while disadvantaging the majority and creating a large number of losers. There are comments that kindergartens are teaching our children a curriculum that is too advanced for their age; school children have to cope with too much homework; and the structure of basic education is fraught with hurdles and dead-ends (p. 15).

As a remedy, education reformers devoted tremendous efforts to alleviating the exam-oriented culture. The Education Commission (2000) called upon teachers, students, parents, and the society in general to "shake off their traditional concept of assessment and embrace the new assessment culture" (p. 70). Beyond conventional exams, Hong Kong policy makers and education reformers had explored alternative assessment models. The goal is no longer achieving high test scores only, but also the acquisition of general and holistic competencies for work and life. Simply put, the focus of Hong Kong education had shifted from performance to competence (Jones & Moore, 1995; Tan, 2019; Tsatsaroni & Evans, 2014). Since the reform, the learning process has become more enjoyable, and the learning objectives are now more diverse. The underlying idea of the reform is "through road", meaning that students can enjoy studying without suffering from the test anxiety caused by public examinations. It was hoped that being relieved from examination pressures, the learners would have more freedom to develop their potentials and engage in more meaningful activities. Specifically, multiple public exams would be reduced to only one—the Hong Kong Diploma of Secondary Education (HKDSE). The Hong Kong Exam Authority shifted the weight of assessment from public exams to internal evaluation mechanism. The primary goal of internal evaluations is to "facilitate learning and teaching and help teachers understand the learning progress and needs of their students" (Education Commission, 2000, p. 10). Specifically, learners would be provided with essential lifelong and life-wide learning activities for project-based evaluation and whole-person development, rather than merely relying on traditional paper-and-pencil tests. In addition, the norm-referenced approach of assessment would be replaced by the criterion-referenced method, meaning that students do not have to compete against each other to be outstanding (Poon & Wong, 2005, 2008).

Furthermore, in the old Secondary School Places Allocation (SSPA) system, students were classified into five bands (i.e., band 1 being the top and band 5 being the bottom), and all Primary 6 students were required to take the Academic Aptitude Test (AAT) for placement among the five bands. Education reformers realized that this banding system had created a labeling effect and that the AAT added immense pressure on students. As discussed in Chap. 1, this banding system was de-emphasized by abandoning the Academic Aptitude Test in 2000, and the gap between elite schools and ordinary schools was narrowed to three bands only, resulting in reduced exam pressure. Nonetheless, some critics have charged that the "through road" approach might cause a decline in academic standards when the function of exams is downplayed (Poon & Wong, 2005, 2008). In any case, because the Hong Kong education system is moving away from an exam-oriented culture, modeling itself after Hong Kong would by no means make the US become more exam-focused. The detail of such reform in math and science education will be discussed next.

2.1.2 Mathematics Education

In terms of mathematics education, HKEDB (2017c) expects that students need knowledge and skills that will help them live a full life in the twenty-first century, the so-called Information Age. It holds that knowledge of mathematics is a necessity for every individual if they are to contribute toward the prosperity of Hong Kong. It bases this claim on the belief that mathematics pervades all aspects of life, and it is not possible to live in the modern world without making some use of mathematics. Thus, many of the developments and decisions made in industry and commerce, the provision of social and community services, as well as government policy and planning, rely to some extent on the use of mathematics.

HKEDB (2017c) maintains that mathematics is essential for the school curriculum of Hong Kong because it is a powerful means of communication, a tool for studying other disciplines, an intellectual endeavor, and a mode of thinking. Indeed, it is a discipline through which students can develop their ability to appreciate the beauty of nature, think logically, and make sound judgments. It further establishes that mathematics is valuable for helping students develop the necessary skills for lifelong learning. It is an integral part of general education and hence an important key learning area (KLA) in the Hong Kong school curriculum.

In terms of the direction of mathematics teaching and learning, HKEDB (2017c) holds that the appropriate use of information technology (IT) in mathematics learning should be emphasized. It considers that high technology items like computers and calculators have profoundly changed the world of math education. It underscores that students need to master IT to adapt to the dynamically changing environment and that with the help of IT tools, meaningless drilling and obsolete topics are neither essential nor relevant in mathematics learning.

HKEDB (2017a) further states that the following objectives are important for Hong Kong students: to acquire the foundations of knowledge and skill; to develop the capabilities for learning how to learn; to think logically, creatively, and critically; to develop and use knowledge; to analyze and solve problems; to access and process information; and to make sound judgments and to communicate with others effectively. Finally, the HKEDB expects students to be enabled to build up confidence and positive attitudes toward mathematics learning, to value mathematics, and to appreciate the beauty of mathematics.

2.1.3 Science Education

In Hong Kong, the lower division of secondary education (S1–S3, equivalent to G7–G9 in the US) uses the Integrated Science curriculum, whereas in the upper-division (S4–S6, equivalent to G10–G12 in the US), science is split into physics, chemistry, and biology. Integrated Science provides students with a broader range and a deeper level of scientific ideas. Rather than merely absorbing information from the teachers and the textbooks, students have many opportunities to reflect on issues and controversies about science and technology, so that they can become science-minded citizens who are capable of making data-driven and informed decisions. Types of reasoning modes for developing and evaluating arguments—such as assessing the reliability of data, the evidence of associations and causal inferences, and the risks and benefits of using certain technologies—help students to be better prepared for solving everyday problems. By broadening their scientific knowledge, the Integrated Science curriculum also helps students to lay a solid foundation for further training or career, including in business administration, law, dental surgery, risk management science, actuarial science, information engineering, and sports science (HKEDB, 2017d).

For the upper-division (S4–S6). HKEDB advised that the learning and teaching activities employed should also aim to promote comprehension, not just the surface-learning of isolated facts. Effective learning is more likely to be achieved when students engage actively rather than observe passively. Additionally, scientific ideas should be open to discussion and negotiation, instead of simply having students learn the material. Further, the content should be learned as a coherent whole, not in a piecemeal fashion (HKEDB, 2015).

2.2 Alignment with the Global Trends: Delors and OECD Reports

It is noteworthy that HKEDB did not develop these schemes alone. Rather, HKEDB looked beyond Hong Kong and made their goals in alignment with global trends. The notion of lifelong learning was incubated and gradually developed by the UNESCO Institute for Education (UIE) during the 1960s. However, at that time some educators who promoted lifelong learning focused on empowering adults though this concept included all stages of education. During the 1970s OECD introduced the idea of recurrent education in an attempt to strengthen formal adult education and vocational training. Unfortunately, both movements failed to yield fruitful results because the strategies employed by most governments were piecemeal and ad hoc. More importantly, lifelong learning was regarded as providing a second chance for at-risk adults or workers who need professional development. As a result, most educators overlooked the importance of lifelong learning to children (Tuijnman & Boström, 2002). Nonetheless, a new paradigm emerged in the 1990s. At that time the international community started to realize the limitation of the traditional approach to lifelong learning. According to Tuijnman and Boström (2002), educators reached a consensus that lifelong learning "embraces all learning that takes place from infancy throughout adult life, in families, schools, vocational training institutions, universities, the workplace, and at large in the community" (p. 103).

To be more specific, in 1996 the United Nations Educational, Scientific, and Cultural Organization released a report entitled *Learning the treasure within: Report to UNESCO of the International Commission on Education for the 21st century* (Delors, 1996). This report, known as the *Delors Report*, proposed a holistic and integrated vision of education based on the overarching principle of lifelong learning, as well as four pillars of learning: to live together, to know, to do, and to be. This report was developed in response to the widespread criticism that educational policies are always driven by economic and financial reasons, neglecting the wellbeing of the learners. The report asserted that very often mere economic growth is achieved at the expense of social equity, respect for the human condition, and respect for natural assets. To rectify the situation, the UN Commission recommended a holistic approach to education, which is based on the four pillars as follows (Delors, 1996):

1. *Learning to live together*: This is about the appreciation of diversity. By understanding other peoples and their history, traditions, and spiritual values, people can make inclusive decisions and ecumenical policies, resulting in fewer conflicts and perpetual peace.

2. *Leaning to know*: Given the rapid changes due to scientific progress and globalization, education should emphasize integration and deep-learning of diverse information across a number of subjects. This general background of various domain knowledge can build momentum for lifelong education.

3. *Learning to do*: This is about applying knowledge to real tasks across a variety of situations. It is impossible for the teacher to exhaust all scenarios in the classroom. Usually, challenges in real life are unpredictable, but today many educational methods do not pay enough attention to the instability and ambiguity of real-life situations. In many cases, applications require pupils to exercise an adaptive mind and take risks, daring to try out different approaches.

4. *Learning to be*: In the twenty-first century, everyone will need to be more independent and make better judgments in order to achieve common goals in society. If we can unleash everyone's talents and encourage everyone to contribute one's own talents to the community, no doubt our overall social wellbeing can be substantively improved. These talents include reasoning power, imagination, physical ability, aesthetic sense, communicative aptitude, and the natural charisma of a group leader.

Besides the *Delors Report*, HKEDB also closely monitors other educational trends in the international community. For Hong Kong educators, the notions of "learning to learn" and "lifelong learning" are tightly tied to research conducted by the Organization for Economic and Cooperation Development (OECD, 2009). To facilitate learning to learn, students must be enabled to actively manage their own learning; they must develop certain self-regulating abilities, such as choosing appropriate learning goals, using prior knowledge to direct their learning, selecting learning strategies that can fit into the task at hand, etc. These skills are considered necessary for building momentum toward lifelong learning (Boekaerts, 2009; Ryan & Deci, 2009; Zimmerman & Cleary, 2009). The key is self-regulation or self-control. A lifelong learner must be an independent learner who is both capable of recognizing what he or she already knows (and does not know) and willing to look for new resources to fill the knowledge gap independently. Take statistics as an example. Currently, the Hong

Kong Curriculum and Assessment Guide of mathematics for Secondary 4–6 (Equivalent to Grade 10–12) includes classical statistics only, but data science is excluded (Hong Kong Examinations and Assessment Authority, 2018). Nonetheless, a student who possesses the mentality of lifelong learning would be able to pursue emerging analytical methods on their own. While doing field observation in Hong Kong, the first author of this book found that secondary school students were enthusiastic to sign up for extra-circular seminars in data science.

Another source that inspired HKEDB to emphasize lifelong learning is the OECD report (2018) entitled "The future of education and skills: Education 2030". In this report OECD introduced the *Learning Framework 2030*, which entails three components constituting competencies:

1. *Creating new values*: "To prepare for 2030, people should be able to think creatively, develop new products and services, new jobs, new processes and methods, new ways of thinking and living, new enterprises, new sectors, new business models and new social models" (OECD, 2018, p. 5). However, these things are hard to achieve if learners cannot take the initiative to constantly reinvent themselves. Thus, teachers must convey the message that knowledge is an open system and that learning is an ongoing process.

2. *Reconciling tensions and dilemmas*: "To be prepared for the future, individuals have to learn to think and act in a more integrated way, taking into account the interconnections and inter-relations between contradictory or incompatible ideas, logics and positions, from both short- and long-term perspectives. In other words, they have to learn to be systems' thinkers" (OECD, 2018, p. 5). Tensions and dilemmas are situational: no two conflicts are exactly the same, and there is no universal or mechanical solution for resolving all conflicts and tensions. Thus, lifelong learners must continue to acquire contextual information in order to customize the solution.

3. *Taking responsibility*: "Dealing with novelty, change, diversity and ambiguity assumes that individuals can think for themselves and work with others. Equally, creativity and problem-solving require the capacity to consider the future consequences of one's actions, to evaluate risk and reward, and to accept accountability for the products of one's work" (OECD, 2018, p. 6). A responsible individual will not use the excuse "I was just following orders" to justify carrying out questionable behavior. A lifelong learner must take responsibility for gathering new information and reflecting upon his or her own actions.

2.3 Learning to Learn in Hong Kong Education Policies

The framework of Hong Kong math and science education was developed in alignment with the *Delors Report* and the OECD report. In the past, Hong Kong followed the British model by treating education as a screening and tracking tool. To be specific, learning abilities were assessed and ranked before the completion of students' secondary school years. Outstanding students were assigned to the track for university education and advanced intellectual inquiry, whereas average students could receive only vocational training (or no further education at all). However, by the end of the twentieth century, Hong Kong's economy faced unprecedented challenges due to advances in technology, and thus the previous elitist model was no longer sustainable. In the past, young people who received only a nine-year education could easily find blue-collar, unskilled jobs in manufacturing plants. However, after 1979, China started reforming its economic structures by adopting an open-door policy to investment. As a result, many labor-intensive jobs had mostly moved into southern China. At the same time, the expansion of Hong Kong's service sector required higher skill levels. This became the driving force for Hong Kong's comprehensive education reform, which began in 1999 and still continues today (OECD, 2012). In the first phase of the post-1997 education reform, about 800 community leaders were invited to a town hall meeting to express their concerns and air their frustration. In 1999 and 2000, the Curriculum Development Institute conducted a thorough review of the school curriculum in order to redesign it to help students meet the challenges of a knowledge-based, interdependent, and changing society, and a globalized world. A year later, the Curriculum Development Council [CDC] (2001) published the groundbreaking document, entitled *Learning to Learn*, which became the overarching theme of the subsequent education reform. Basically, the focus of education would shift from "teaching" to "learning", meaning that learning would become a constructive process through active engagement by the learners. In addition, learning is about comprehension, which should be demonstrated by integration and application of the knowledge to real-life situations, rather than memorization. The philosophical belief behind this new approach is that all students have the ability to learn, and thus everyone should be enabled to engage in lifelong learning by educational institutions offering essential learning experiences. More importantly, curriculum development should be an ongoing process so that it can adapt. In lay terms, learning to learn is characterized by the following features (Bransford, 2000; Sawyer, 2006):

- Learning is the active construction of knowledge by the learner.
- Learning is a process, achieved through activities called learning experiences.
- Similar experiences may lead to the construction of different kinds of knowledge, i.e., to people learning differently.
- Learning is for understanding.
- Understanding is demonstrated by the effective application of the knowledge thus constructed.
- Effective learning experiences often require the integration of knowledge.
- Learning is, therefore, best in real-life experiences with actual effects.
- Learning is also a social action, best achieved in groups.
- Human learning is motivated by a sense of improvement.

Besides emphasizing creativity, critical thinking, and problem-solving, the new curriculum also aims to help students to build a value system and shape their positive attitudes. Values are personality qualities that students should develop for making decisions, whereas attitudes are personal dispositions for performing tasks well. Hence, this paradigm shift placed more emphasis on preparing students across a spectrum of abilities through multiple pathways (CDC, 2001). After years of discussion, design, experimentation, and evaluation, the new curriculum was launched in 2009 (OECD, 2012).

As mentioned before, Hong Kong curriculum development is continuous and adaptive. The Basic Education Curriculum Guide (Primary 1–6, equivalent to Grades 1–6 in the US) was updated in mid-2014. In the 2014 update, the CDC used the term "Learning to Learn 2.0" to emphasize that curriculum development had coped with social changes since the introduction of the first version of Learning to Learn in 2001 (CDC, 2015). After going through a comprehensive process of data collection and evaluation, in 2017 Hong Kong Education Bureau (2017a, b) affirmed the value of Learning to Learn 2.0. The following renewed or updated goals, which are related to science and math education, were proposed in order to cope with social trends:

- Promoting STEM education and ICT in education
- Fostering an entrepreneurial spirit
- Diversifying life-wide learning experiences
- Stepping up gifted education.

During the 2017 update, it was discovered that over the previous decade, Hong Kong schools had made good progress due to the curriculum reform. Many schools have developed innovative curricula, pedagogical strategies, and rich resources to enhance student learning. With the adoption of the student-centric approach, a wide range of learning experiences were provided to achieve whole-person development and to build momentum for continuous learning beyond graduation. As a result, students have greater learning agility, stronger transferable skills, and are more proactive. Teachers have successfully adopted a paradigm shift from teacher-centric to student-centric pedagogical approaches, and they are willing to share resources and collaborate with each other. Students have opportunities to actively participate in learning. The 2017 evaluation also pointed out that lifelong learning is especially important in high-tech environments (Hong Kong Examinations and Assessment Authority, 2016; CDC, 2015). Two decades ago, one needed to know how to configure a modem to get online, but today basic knowledge of Wi-Fi networks is essential for performing day-to-day tasks. Thus, in 2007 the CDC asserted that in order to prepare students for the challenges in a knowledge-based and technologically advanced world, it is necessary to promote self directed learning in school curriculum development so that learners can acquire a sense of control and metacognitive skills (CDC, 2015). In conclusion, HKEDB (2017a, b) asserted that learning to learn will continue in the ensuing years leading toward 2030 and beyond, which echoes the OECD Learning Framework 2030 (OECD, 2018). In 2018 HKEDB issued another update, which incorporates the global trends found in five international organizations and 24 countries or regions. Besides re-asserting the importance of problem-solving, creativity, critical thinking, learning to learn, and lifelong learning, this update also emphasized communication, collaboration, self-regulation, civil responsibility, and social participation (HKEDB, 2018).

At the time of this writing, the Hong Kong political situation makes one of the preceding learning goals, critical thinking, more relevant than ever. Perhaps the 2019 protests in Hong Kong may be a manifestation of the value of education for critical thinking. Starting from June 2019, many Hong Kong people, including university and secondary school students, have protested against the proposal that allows extradition of Hong Kong citizens to Mainland China. In September, former Chief Executive of Hong Kong Tung Chee-hwa and pro-Beijing supporters blamed the social turbulence on the failure of education (May & Qin, 2019). On the contrary, it is the firm conviction of the authors that reflections and debates about the merits of democracy and civil rights are the desirable outcomes of education for critical thinking, though violence against innocent people should be condemned. Some educators even grabbed this as a golden opportunity to train children in critical thinking with regard to logical communication (Wong & Wong, 2019) and the rule of law (Sieh, 2019). Even before the protests, there was a growing concern that Beijing's role presented challenges to Hong Kong education because Mainland China urges Hong Kong educators to introduce patriotism into Chinese history lessons and other subjects (Forestier, 2018). Nonetheless, the focus of this book is on math and science education and thus is it beyond the scope of this project to discuss whether the subject of Chinese history would be turned into a tool of brainwashing.

2.4 Lifelong Learning in Practice

2.4.1 Learning by Teaching

The notion of lifelong learning may seem idealistic, but how can Hong Kong teachers translate this abstract idea into actionable items? One of the pedagogical methods for achieving this goal is to make students become teachers. Interestingly, a good teacher tends to be a lifelong learner because this profession demands him or her to stay current with the field. To be more specific, a teacher must fully understand the concepts to be taught; otherwise, he or she might be embarrassed by students who raise deeper questions about the subject matter. In addition, it is imperative for the teacher to update his or her knowledge continuously. Needless to say, today every discipline is changing rapidly. Two decades ago, physics majors did not learn much about neutrino, Higgs boson, dark matter, and dark energy; likewise, biology majors heard nothing about stem cell research and genetic engineering. By the same token, today in computer science the hottest programming languages are Java, R, and Python. COBOL and FORTRAN are nothing more than historical relics.

The philosophy of learning by teaching is epitomized by the following saying attributed to Harbhajan Singh Yogi: "If you want to learn something, read about it. If you want to understand something, write about it. If you want to master something, teach it". Hence, Hong Kong science teachers attempted to motivate secondary school students to be lifelong learners by launching the *Gopher Science Labs* (GSL), a peer-to peer-learning program in which secondary school students teach a range of science activities to primary school students (Grobler, 2016a). The labs urged students to employ experimental methods to answer various research questions, such as "Why do slugs and snails need slime?" "Do you see what I see?" "Can you beat the test?" "The Lemon Battery" and "Don't blow it!" and "Why do leaves change color in autumn?" All experiments were taught by the participating secondary school students, with teachers merely playing the role of facilitation.

Based on the feedback from the student-teachers, it was believed that this approach has been effective in motivating students to learn more in science when they realized their knowledge gap during teaching preparation. For example, a student said,

> Before I participated in this event, I thought that preparing materials to teach primary school students would be easy since the steps and principle of the experiment are written on the papers given to us and we only had to follow them. However, when my group mates and I were preparing for this event, I understood why people always appreciate the hard work of teachers. We spent more than 3 hours testing the experiment we were going to teach, which has only 8 steps, not to mention the preparation of the scripts and the PowerPoint (Grobler, 2016b, p. 81).

Another student wrote, "By teaching the students, I have also learnt the theory behind the experiment. I was responsible for the soap powered boat experiment and we had to prepare all the materials and the script. The soap powered boat experiment is a simple experiment which the primary school students can try at home and I am happy that they liked the experiment and were amazed by it" (Grobler, 2016b, p. 82).

2.4.2 Science as a Mentality

It is noteworthy that inspired by the guideline "Learning to learn" issued by the Hong Kong Education Bureau, many Hong Kong educators view math and science learning as developing a mentality or an attitude, rather than following a set of rules and formulas. Specifically, Wong (2018) developed a problem-solving flow chart for science and math education, as shown in Fig. 2.1. The underlying philosophy of this approach is that scientific inquiry and problem-solving are interdisciplinary in nature, and thus educators should not confine the learner's mind into a canned domain or a preset path. In the past, mathematicians, scientists, and engineers could deal with problems in isolation because their problems were not complicated or integrative in nature. However, today if a person lacks an integrative mind and is merely a specialist, needless to say, his or her problem-solving skills will be limited. To put it another way: in the real world, there are no physics-only, chemistry-only, or computer-only problems. It is imperative to train the student to look at problems from different perspectives, and to solve problems using all available resources.

The upper left of the graph shows that the student starts with a sense of wonder and a desire to understand natural phenomena; the upper right indicates the importance of awareness of the fact that every day we face problems calling for practical solutions. When the learner approaches the phenomenon or the problem, he or she should not be given a modal answer or a cookbook-like step-by-step guideline. In the real world, a problem is inherently ill-defined and ambiguous. The student needs

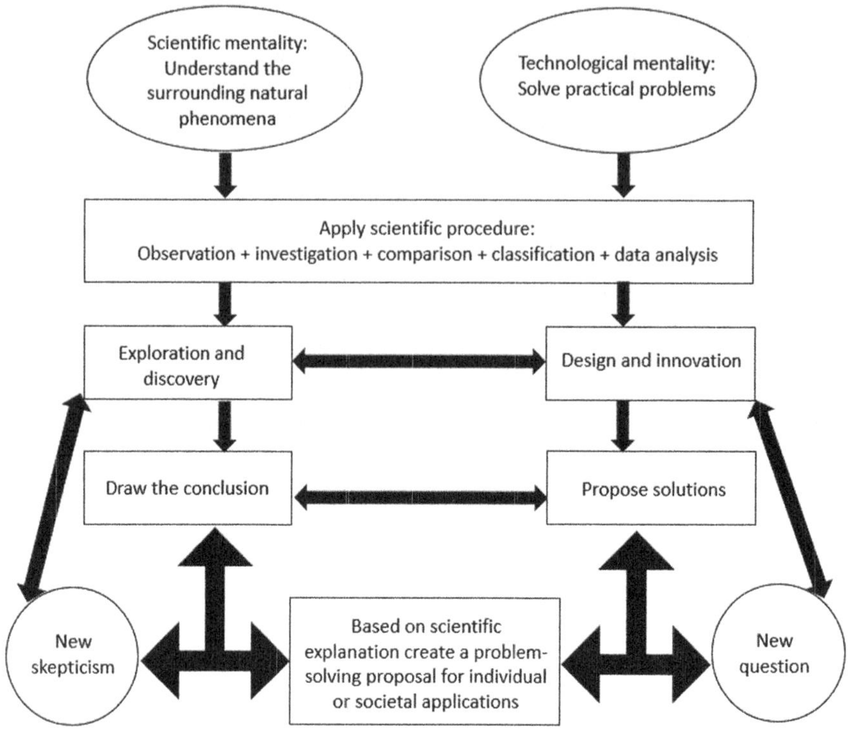

Fig. 2.1 Science as a mentality for problem-solving

to employ scientific protocol, such as observation, investigation, classification, comparison, and data analysis in order to thoroughly explore the data and to develop an innovative solution. This step might yield several working hypotheses or conjectures. Nonetheless, these initial conclusions or proposed solutions are not final. Rather, this inquiry is an iterative process, which necessitates ongoing questioning and evaluation. In the end, based upon scientific explanation, the learner develops the best inference and actionable items for an individual or societal applications.

The purpose of science and math education should not be narrowed down to training elites to be professional scientists, engineers, and computer programmers. In the past, this approach only benefited a few elites, but the majority was alienated. Rather, math and science education should be relevant to everyday life. A broader objective is to educate the whole generation to be "scientific citizens" that have a scientific mind for solving any problem, regardless of the nature of the problem encountered (Wong, 2018).

2.4.3 Teaching Nature of Science (NOS)

Teaching science as a mentality or a way of life is fundamentally different from the conventional approach that emphasizes equations and formulas in a piecemeal fashion. When science is viewed as a way of life or a philosophical perspective, one must learn the nature of science (NOS) or the essence of scientific methods by looking at the "forest" instead of the "trees" (Abd-El-Khalick & Akerson, 2004). The notion of NOS as a crucial component of scientific literacy can be traced back to the 1960s and 1970s. During that period of time, a group of Harvard scholars shifted the focus of inquiry from seeking scientific knowledge to looking for knowledge of science or meta-science (Duschl, & Grandy, 2013; Holton, Rutherford, & Watson 1970). Since then, this idea has been drawing attention from US educators; indeed, several education policy documents were built on this idea, such as *Science for all Americans* (American Association for the Advancement of Science, 1989) and the *Science Education Content Standards* (National Research Council, 1996). Since the 1990s, the pedagogical approach of teaching NOS has spread to other nations, including Australia, Canada, Colombia, the United Kingdom, Ireland, New Zealand, Indonesia, Lebanon, Mexico, Thailand, and South Africa, but NOS tenets vary from country to country (Erduran & Dagher, 2014; McComas & Olson, 1998; Olson, 2018). However, Lederman and Lederman (2014) found that the importance of NOS has been de-emphasized in US education documents since the late 1990s. This trend is not surprising because as a decentralized system, US schools do not have to treat NOS as a mandatory component in science education, and thus NOS guidelines are more advisory than prescriptive (Olson, 2018). In the US only 6% of science teachers taught some content related to NOS (Backhus & Thompson, 2006), but this figure might be misleading because different teachers might interpret NOS differently and some might not explicitly refer to their curriculum as NOS, especially when there is no consensus view regarding NOS (McComas, 2002).

Hong Kong's education system, as a top-down hierarchy, makes NOS ubiquitous. For Hong Kong educators, teaching NOS can stir up learner curiosity, resulting in building momentum for ongoing inquiry and lifelong learning. Hence, many Hong Kong teachers explicitly include NOS into their curriculum (Liu, 2016). The Hong Kong version of NOS is in alignment with the view of Duschl and Grandy (2013): NOS is about an inquiry of inquiry, which includes exploration, formulation, measurement, observation, argument from evidence, model building, and interpretation. Duschl and Grandy (2013) explain, "Students should learn through experience what it means to be rational and objective, and not to simply accept those adjectives as descriptors of science" (p. 2112). During the inquiry of NOS, Hong Kong students are asked to contemplate on the following questions (see Table 2.1):

When the students are assigned to conduct an experiment to answer a research question, not only do they need to answer the preceding questions but also they need to choose the materials and instruments that are appropriate for the nature of the research question. The following are two examples:

Table 2.1 Open-ended questions for inquiry of nature of science

	NOS elements	Questions
1	Empirical-oriented	What are the criteria for a scientist to judge whether an assertion is correct?
2	Creativity	Do you think that your experiment is creative? Please explain
3	Rigorous	Why do you need to take variable manipulation and assumptions into consideration in an experiment?
4	Replicability	Why do you need to document the detail of the experimental procedures and protocols? Please explain
5	Credibility	Why do different researchers need to verify the scientific experiment? If it cannot be verified, what are the problems?
6	Limitations and uncertainty	Why are there errors? Please explain
7	Tentativeness	Do you think your conclusion is resulted from observation or deduction? Please explain
8	Skepticism	Do you think that your conclusion can fit the reality? Please explain

Experiment 1
After the skin of an apple is peeled off, the surface of the apple turns from white to yellow. Conduct an experiment to find a way to preserve the original color of the apple. Choose your materials and instruments from the following list:
Given: An apple

Experiment 2
When you cut an onion, your eyes are irritated. Conduct an experiment to find a way of cutting an onion without irritation. Choose your materials and instruments from the following list:

Given: An onion
In both experiments, the students are given the following options and are asked to design the experiment based on their choices. They have to choose the appropriate apparatus and also specify the number of each item needed for the experiment (see Table 2.2).

Teachers who employed this pedagogical approach found that learners could better understand the nature of science (Liu, 2016). Regardless of whether these students become scientists in the future, this science literacy or scientific attitude will benefit them in all aspects of their life.

2.4.4 Lifelong Learning and Science as an Attitude

In 1996, OECD had reached the consensus that promoting lifelong learning is a priority for its member nations (OECD, 1996; Tuijnman & Boström, 2002). Four years later, Hong Kong Education Commission (2000) adopted this suggestion in

Table 2.2 Items available for experimentation and inquiry of NOS

Item number	Item name	Quantity	Item number	Item name	Quantity
1	Test tube		13	100 mL measuring cylinder	
2	Test tube rack		14	Glass stick	
3	Label		15	Gauze	
4	Plastic wrap		16	Petri dish	
5	Cork drill		17	Wax paper	
6	Knife		18	Blue litmus paper	
7	Tweezers		19	Red litmus paper	
8	White tile		20	pH test paper	
9	Filter		21	Candle	
10	Plastic head dropper		22	Match	
11	100 mL beaker		23	Mortar	
12	250 mL beaker		24	Tin foil	

an attempt to migrate the focus from elitism to training all learners with science literacy. This new orientation is based on the noble ideal of fairness and social equality. In the past, only a few high-performing students had the opportunities to receive postsecondary education, becoming professionals upon graduation, but the majority were stuck in the lower level of the social ladder. To rectify this situation, from that point on all learners would be equipped with the tool of ongoing learning so that they could still be successful in pursuing alternate pathways (Kennedy, 2011).

A teacher interviewed by the team shared his specific strategies for promoting lifelong learning. Basically, the content of the subject matter is treated as a means to an end; otherwise, it seems useless when the students do not take math or science as their majors in the future. Science literacy, put bluntly, becomes a way of life, or a way of ongoing learning to acquire information in different domains. One of his emphases in teaching math and science is concerned with how students approach complex challenges. Another concern is how students approach their changing environment. By achieving the second goal, learners are urged by the teacher to develop the mentalities of curiosity, taking initiative, persistence, adaptability, leadership, and social/cultural awareness. The core message is: the only constant in life is change. If you cannot cope with the changing environment, you will be defeated by your rivals who are willing to use cutting-edge methods. To actualize this idea, this teacher introduced a science camp program in which students are asked to redesign an outdated shopping mall located in Kwai Chung, the northwestern region of Hong Kong. He explicitly said, "The camp is not to develop students' academic knowledge and skills, but instead, more on attitude development". Three decades ago the shopping mall was state-of-the-art, but today everything is run down, and thus many shops are facing difficulties in attracting customers. In the camp, the learners were asked to apply their scientific knowledge toward revitalizing this dying mall.

2.4.5 *Scientific Discourse*

In light of treating science as a mentality or a lifestyle, teachers can motivate learners through "scientific discourse". During the discourse, the teacher might throw out interesting questions to the learners, such as "How old is the universe?" "How big is the universe?" "Are there alien civilizations in the universe?" The learners could easily obtain different answers using an Internet search engine. However, different sources might offer different answers, and this becomes a teachable moment. Instead of telling the students which answer is correct or is more likely to be right, the teacher should encourage the students to keep an open mind to all different answers, thus leading to comparison and further analysis.

Alternately, the teacher could use more practical problems for students who are more career-oriented. For example, "If Hong Kong needs to build an additional cross-harbor tunnel (i.e., another bridge connecting Hong Kong Island and Lantau Island), where is the best location? What are the potential problems in engineering and construction? How can we solve these problems?"

By the same token, mathematics can be taught by challenging students with practical problems. Calculation is only a small portion of math; the vast majority is about logical reasoning and conceptualization. The teacher might use a problem like the following: "You have a stack of magazines. What is the best way to categorize them?" Some might sort them by date, some might classify them by publisher, and some might use the subject matter as the criterion. Mathematically speaking, the sorter should follow the principles of mutual exclusion and collective exhaustion. Simply put, under the categorization each item should belong to one class only, and the result should be informative to the user, meaning that he or she can easily retrieve the item when it is needed (Wong, 2018).

2.5 Conclusion

The authors of this book have taught math and science subjects at different levels in the US. Very often we have faced strong resistance from the students because they did not see the reason for learning math and science, especially when their career goals were unrelated to science and math. As a remedy, learning to learn aims to redirect science and math education from the old paradigm of training specialists toward a new direction that is consonant and compatible with the emerging demands of science and math in the present and the future. The key point is that not every learner should become a scientist or an engineer in the future. Rather, when most people become scientific citizens who are proficient in science literacy, this mentality, attitude, or lifestyle will benefit them personally and the entire society more broadly. When scientific inquiry becomes a lifestyle, it necessitates and facilitates lifelong learning.

It is important to point out that learning to learn is not a new idea developed in Hong Kong. Rather, HKEDB humbly borrowed the notion from the international community. What distinguishes the Hong Kong approach is that on the one hand Hong Kong educators did not transplant policies and practices from the West without regard to cultural differences, but on the other hand they did not use context as an excuse for not adopting practices that had worked well elsewhere. Hong Kong students and teachers skillfully blend Chinese traditional and Western progressive approaches, such as memorization of basic information and comprehension of advanced concepts; collaboration and competition; and didactic and constructivist models (Chan & Rao, 2009). This integration is based on the "doctrine of the mean" in Confucianism, which emphasizes accommodation and compromise rather than leaning on the extremes (Forestier & Crossley, 2015).

Hong Kong will continue to draw reference from Eastern and Western pedagogical approaches and cultures. In an obvious manner, Hong Kong secondary school teachers show confidence in the education system. One teacher interviewed by the team commented, "Hong Kong has had a strong and well-established education system in math and science for many years. It is proven to train up students very effectively and efficiently". This small region far to the east of the US has dreamed big in constructing its educational platform that will leverage its competence on a world-class level. The underlying suspicion of ideas developed elsewhere may operate within an understandable logic, that each nation has embraced different educational beliefs, history, socioeconomic system, and demographics. However, if the UK adamantly holds onto such suspicion in the face of the evident growth and success of Hong Kong's education reform, it will leave the UK stagnant in this dynamically changing world. Fortunately, the UK has recognized the merits of Hong Kong in incorporating the strengths of Western educational philosophy into its traditional form. To the British people, learning from Hong Kong is a "two-way borrowing" (Forestier & Crossley, 2015, p. 678). It is the conviction of the authors that the US education system can also benefit from this type of integration and two-way borrowing.

References

American Association for the Advancement of Science. (1989). *Science for all Americans*. Washington, DC: Author.

Abd-El-Khalick, F., & Akerson, V. (2004). Learning as conceptual change: Factors mediating the development of preservice teachers' views of nature of science. *Science Education, 88*, 785–810.

Backhus, D. A., & Thompson, K. W. (2006). Addressing the nature of science in preservice science teacher preparation programs: science educator perceptions. *Journal of Science Teacher Education, 17*(1), 65–81.

Boekaerts, M. (2009). Goal-directed behaviors in the classroom. In K. R. Wentzel, & A. Wigfield Chan (Eds.), *Handbook of motivation at school* (pp. 105–122). New York, NY: Routledge.

Bransford, J. D. (Ed.). (2000). *How people learn: Brain, mind, experience, and school*. Washington, DC: National Academy Press.

Chan, K. K., & Rao, N. (2009). Moving beyond paradoxes: Understanding Chinese learners and their teachers. In C. K. K. Chan, & N. Rao (Eds.), *Revisiting the Chinese learner: Changing contexts, changing education: CERC Studies in Comparative Education* (Vol. 25, pp. 3–34). Hong Kong: Springer/Comparative Education Research Centre, The University of Hong Kong.

Curriculum Development Council. (2001). *Learning to learn-The way forward in curriculum development*. Hong Kong: Author. Retrieved from https://www.edb.gov.hk/mobile/en/curriculum-development/cs-curriculum-doc-report/wf-in-cur/index.html.

Curriculum Development Council. (2015). *Promotion of STEM education: Unleashing potential in innovation*. Hong Kong: Hong Kong Administrative Region of the People's Republic of China. Retrieved from http://www.edb.gov.hk/attachment/en/curriculum-development/renewal/Brief%20on%20STEM%20(Overview)_eng_20151105.pdf.

Delors, J., et al. (1996). *Learning: The treasure within*. Paris: UNESCO. Retrieved from http://unesdoc.unesco.org/images/0010/001095/109590eo.pdf.

Duncan, A. (2010). The vision of education reform in the United States. *US Department of Education*. Retrieved from https://www.ed.gov/news/speeches/vision-education-reform-united-states-secretary-arne-duncans-remarks-united-nations-ed.

Duncan, A. (2011). Lessons from high-performing countries. *US Department of Education*. Retrieved from https://www.ed.gov/news/speeches/lessons-high-performing-countries.

Duschl, R. A., & Grandy, R. (2013). Two views about explicitly teaching nature of science. *Science & Education, 22,* 2109–2139. https://doi.org/10.1007/s11191-012-9539-4.

Erduran, S., & Dagher, Z. R. (2014). Regaining focus in Irish junior cycle science: Potential new directions for curriculum and assessment development on nature of science. *Irish Educational Studies, 33,* 335–350.

Forestier, K. (2018). Hong Kong education reform continues, but Beijing's role presents new challenges. *South China Morning Post*. Retrieved from https://www.scmp.com/comment/insight-opinion/article/2131727/hong-kong-education-reform-continues-beijings-role-presents.

Forestier, K., & Crossley, M. (2015). International education policy transfer–borrowing both ways: the Hong Kong and England experience. *Compare: A Journal of Comparative and International Education, 45*, 664–685, https://doi.org/10.1080/03057925.2014.928508.

Gove, M. (2013). Written statement to Parliament: Education Secretary Michael Gove sets out plans for the new national curriculum. Retrieved from http://www.gov.uk/government/speeches/education-reform-schools.

Grobler, G. C. (2016a). Interdisciplinary forensic science programme. *Hong Kong Science Teachers' Journal, 31,* 14–18.

Grobler, G. C. (2016b). Let our students teach! Teaching as a way of mastering knowledge. *Hong Kong Teacher's Journal, 32,* 79–83.

Holton, G., Rutherford, J., & Watson, F. (1970). *Project physics: Handbook*. Cambridge, MA: Project Physics.

Hong Kong Education Bureau. (2015). *Science education key learning area: Physics curriculum and assessment guide (Secondary 4–6)*. Retrieved from https://www.edb.gov.hk/attachment/en/curriculum-development/kla/science-edu/Phy_C&A_Guide_updated_e_20151126.pdf.

Hong Kong Education Bureau. (2017a). Secondary education curriculum guide: Ongoing renewal of the school curriculum. Hong Kong: Author.

Hong Kong Education Bureau. (2017b). Secondary education curriculum guide: Information technology for interactive learning: Towards self-directed learning. Hong Kong: Author.

Hong Kong Education Bureau. (2017c). *Mathematics education*. Retrieved from https://www.edb.gov.hk/en/curriculum-development/kla/ma/index.html.

Hong Kong Education Bureau. (2017d). *Supplement to the science education key learning area curriculum guide*. Retrieved from https://www.edb.gov.hk/attachment/en/curriculum-development/kla/science-edu/IS_C&A_Guide_updated_e_151126.pdf.

Hong Kong Education Bureau. (2018). *Ongoing renewal of the school curriculum*. Retrieved from https://www.edb.gov.hk/en/curriculum-development/renewal/index.html.

Hong Kong Education Commission. (1999). Review of education system: Framework for education reform. Hong Kong: Author.
Hong Kong Education Commission. (2000). *Learning for life: Reform proposals for the education system in Hong Kong*. Hong Kong: Author. Retrieved from https://www.e-c.edu.hk/doc/en/publications_and_related_documents/education_reform/Edu-reform-eng.pdf.
Hong Kong Examinations and Assessment Authority. (2016). *Category A: Senior secondary subjects*. Retrieved from http://www.hkeaa.edu.hk/en/hkdse/assessment/subject_information/category_a_subjects/hkdse_subj.html?A1&1&4.
Hong Kong Examinations and Assessment Authority. (2018). Assessment Framework for 2021 HKDSE. Retrieved from http://www.hkeaa.edu.hk/en/hkdse/hkdse_subj.html?A1&1&4_1.
Jones, L., & Moore, R. (1995). Appropriating competence: The competency movement, the new right and the 'culture change' project. *British Journal of Education and Work, 8*(2), 78–92. https://doi.org/10.1080/0269000950080206.
Kennedy, K. J. (2011). *Changing schools for changing times: New directions for the school curriculum in Hong Kong*. Hong Kong: Chinese University Press.
Lederman, N. G., & Lederman, J. S. (2014). Is nature of science going, going, going, gone? *Journal of Science Teacher Education, 25,* 235–238.
Liu, C. K. (2016). The positive impact of teaching nature of science on students via scientific experiments. *Hong Kong Science Teachers Journal, 32,* 84–119.
Martens, K., & Niemann, D. (2013). When do numbers count? The differential impact of the PISA rating and ranking on education policy in Germany and the US. *German Politics, 22,* 314–332. https://doi.org/10.1080/09644008.2013.794455.
May, T., & Qin, A. (2019). The high school course Beijing accuses of radicalizing Hong Kong. *New York Times*. Retrieved from https://www.nytimes.com/2019/09/01/world/asia/hong-kong-protests-education-china.html.
McComas, W. F., & Olson, J. K. (1998). The nature of science in international science education standards documents. In W. F. McComas (Ed.), *The nature of science in science education: Rationales and strategies* (pp. 41–52). The Netherlands: Kluwer.
McComas, W. F. (2002). *The nature of science in science education: Rationales and strategies*. Netherlands: Springer.
National Research Council. (1996). *National science education standards*. Washington, DC: National Academies.
Olson, J. K. (2018). The inclusion of the Nature of Science in nine recent international science education standards documents. *Science & Education, 27,* 637–660. https://doi.org/10.1007/s11191-018-9993-8.
Organization for Economic and Cooperation Development [OECD]. (1996). *Lifelong learning for all: Meeting of the education committee at ministerial level*. Paris: Author.
Organization for Economic and Cooperation Development [OECD]. (2009). PISA 2009 results: Learning to learn: Student engagement, strategies, and practices. Paris: Author. Retrieved from https://www.oecd.org/pisa/pisaproducts/48852630.pdf.
Organization for Economic Cooperation and Development [OECD]. (2012). Lessons from PISA for the United States; Shanghai and Hong Kong. Paris: Author. http://dx.doi.org/10.1787/9789264096660-5-en.
Organization for Economic and Cooperation Development [OECD]. (2018). *The future of education and skills: Education 2030*. Paris; Author. Retrieved from https://www.oecd.org/education/2030/E2030%20Position%20Paper%20(05.04.2018).pdf.
Poon, A., & Wong, Y. C. (2005). Equity vs excellence: A preliminary exploration of Hong Kong's education reform. *Centre for Asian Pacific Studies*. Retrieved from https://pdfs.semanticscholar.org/a0b0/ed0cca549370ecb3b040539fedf9c7c20dc5.pdf.
Poon, A. Y. K., & Wong, Y.-C. (2008). Education reform in Hong Kong: The "through-road" model and its societal consequences. *International Review of Education/Internationale Zeitschrift Für Erziehungswissenschaft, 54*(1), 33–55. https://doi.org/10.1007/s11159-007-9073-9.

Ryan, R. M., & Deci, E. L. (2009). Promoting self-determined school engagement: Motivation, learning, and well-being. In K. R. Wentzel & A. Wigfield (Eds.), *Handbook of motivation in school* (pp. 71–196). New York, NY: Taylor Francis.

Sieh, M. (2019). Hong Kong school course promotes critical thinking on rule of law. *South China Morning Post*. Retrieved from https://www.scmp.com/lifestyle/families/article/1826831/secondary-school-course-promotes-critical-thinking-rule-law.

Sawyer, K. (Ed.). (2006). *The Cambridge handbook of the learning sciences*. Cambridge, UK: Cambridge University Press.

Tan, C. (2019). Competence or performance? A Bernsteinian analysis of basic competency assessment in Hong Kong. *British Journal of Educational Studies, 67*, 235–250. https://doi.org/10.1080/00071005.2017.1423461.

Tsatsaroni, A., & Evans, J. (2014). Adult numeracy and the totally pedagogised society: PIAAC and other international surveys in the context of global educational policy on lifelong learning. *Educational Studies in Mathematics, 87*(2), 167–186. https://doi.org/10.1007/s10649-013-9470-x.

Tuijnman, A., & Boström, A. K. (2002). Changing notions of lifelong education and lifelong learning. *International Review of Education, 48*, 93–109. https://doi.org/10.1023/A:1015601909731.

Wiseman, A. (2015). *How the world learns: Comparative educational systems*. Chantilly, VA: The Teaching Company.

Wong, P. W. (2018). *STEM for parents*. Hong Kong: Shing Tao Publisher.

Wong, M. W. K, & Wong, C. H. (2019). Extradition protests a chance for Hong Kong schools to teach critical thinking and polite debate. *South China Morning Post*. Retrieved from https://www.scmp.com/comment/letters/article/3014911/extradition-protests-chance-hong-kong-schools-teach-critical.

Zimmerman, B. J., & Cleary, T. J. (2009). Motives to self-regulate learning: A social cognitive account. In K. R. Wentzel & A. Wigfield (Eds.), *Handbook of motivation in school* (pp. 1247–1264). New York, NY: Taylor Francis.

Chapter 3
Creativity, Work Ethic, and Foundational Knowledge

In 2012, President Obama emphasized that twenty-first century education should prepare students to become innovative. Thus, American students should not merely memorize information, but form novel thoughts. By associating creativity with entrepreneurship, Obama suggested that novel ideas could result in start-ups and new products, thus helping the American economy in the long run (Katz-Buonincontro, 2012). This highlights one common counterargument against adopting the East Asian approach: that the flexible US education system breeds innovation, which is aligned with the national goal proposed by the Obama administration. On the other hand, the East Asian approach may kill creativity among students with its rigid, exam-oriented method. This chapter aims to reexamine this argument. The authors argue that innovative ideas must be built upon a solid foundation of knowledge, as well as certain personality traits, such as persistence and tolerance of ambiguity. Prior research has found that highly creative people devote at least 10 years to fine-tuning their skills before becoming truly productive (Edelson, 1999). In brief, rigorous training and creativity must go hand in hand.

3.1 American Innovation and US Education

Undoubtedly, the US has an impressive history of innovation: airplanes, the assembly line, the atomic bomb, transistors, lasers, microprocessors, etc. However, whether we can attribute US innovation totally to US education awaits further investigation. Interestingly, many legendary US innovators have been dissatisfied with the formal education system, including Bill Gates (co-founder of Microsoft), Steve Jobs (co-founder of Apple, Inc.), and Mark Zuckerberg (co-founder of Facebook). In fact, they dropped out of their universities in order to pursue their dreams. In 1950, J. P. Guilford began his inaugural speech at the American Psychological Association by asking this question: "Why were schools not producing more creative persons? Why is there so little apparent correlation between education and creative productiveness?" (as cited

C. H. Yu and H. S. Lee, *Creating Change to Improve Science and Mathematics Education*, https://doi.org/10.1007/978-981-15-3156-9_3

in Fasko, 2001, p. 444). Dacey and Lennon (1998) bluntly claimed, "Schools suppress creativity" (p. 69). According to Gardner (1991), young children aged 2–7 have a golden opportunity to develop creativity due to their curious nature. However, after receiving formal education, they tend to become more cautious and less creative. Moreover, school culture favors conformity rather than creativity. As a result, students became passive observers rather than active explorers because nonconformity at school could result in punishment by the teacher. Paradoxically, Dacey and Lennon (1998) found that the longer children stay in school, the lower their creativity scores are. By the same token, world-renowned educator and expert on creativity Sir Ken Robinson and his associates (Robinson, 2006, 2011, 2013; Robinson & Aronica, 2015) have emphasized that children have natural talents in the sense that creativity is already embedded in their minds. They are naturally curious, too. Thus, if educators can light up the curiosity inside a child's mind, they can develop the momentum to learn. However, much of the present education system in the US, especially the practices codified under the *No Child Left behind* Act, promotes compliance, conformity, and standardization, rather than fostering creativity. While Hong Kong has only four mandatory standardized exams (Territory-wide System Assessment [TSA] at Grade 3, Grade 6, and Grade 9 [S3], and Hong Kong Diploma of Secondary Education (HKDSE) Examination at Grade 12 [S6]), the average US public-school student takes 112 mandatory standardized tests from preschool to high school; a typical student spends about 25 hours per year taking these mandatory exams, according to the Council of the Great City schools (2015). The Council criticized that these assessments are "not strategic, often redundant" (p. 1). In a similar tone, Puccio (2017) said that standardization leads our education in a direction against creativity, which is a "perfect storm" (p. 334).

Furthermore, although Egan (1988, 2005) saw the value of imagination in creative education, he found that imagination does not play a significant role in the US educational system. A survey conducted by *Time* indicated that 31% of respondents blamed the education system for not cultivating creativity in students (Kluger, 2013). One of the strategies for promoting imagination and creativity is to facilitate the critical evaluation of ideas (Robinson, 2011). However, this critical element is weak or even missing in US schools because the current learning culture skews towards boosting self-efficacy, thus downplaying critical evaluation (see Chap. 5: Self-efficacy and productive failure). Another detrimental element against creativity is over-justification (Lepper, Greene, & Nisbet, 1983). Specifically, when too many rewards, such as tangible payments and praises, are given to the learners, creativity is unintentionally dampened. These rewards could turn the student's goal from learning to getting a reward or high score while taking the least amount of risk (see Chap. 5: Self-efficacy and productive failure).

Today, there are still diverse views about creativity. Most researchers in cognitive science tend to agree that no single mental process can be identified with creative thinking (Dacey & Lennon, 1998). Treffinger (1996) found that there are more than 100 different definitions of creativity and as many assessment tools to evaluate it. Similarly, Soh (2019) found that 162 creativity tests were published between 1994 and 2004, and by now the list would have been extended. Some researchers equate creativity with intelligence, whereas others regard creativity as one of many components of intelligence. Conversely, some treat intelligence as a part of creativity, while others consider them two distinct mentalities (Kaufman & Baer, 2006; Sternberg, 1999). Generally speaking, researchers have accepted that no single ability or trait is the key to creativity. Indeed, creativity involves an interaction of cognitive (information processing), conative (personality traits, motivational aspects), and emotional factors (affective states) (Runco, 2004). If there exists such diverse views on the very nature of creativity, then it is problematic to attribute US innovation to the US education system.

3.2 Personality Traits and Work Ethic

The authors of this book concur with Torrance's (1988) and Weisberg's (1999, 2006) view that creativity is the ability to apply knowledge to a new situation. If there is no knowledge, there will be no creativity. On several fronts, numerous research has sought the true meaning of creativity. The most apparent element that composes creativity is originality, as creativity is often associated with novelty and uniqueness. Yet many agree that the intention and the outcome of this originality must be useful, which leads to the belief that creativity is both original and effective (Runco, 2004, 2014). Stein (1953) neatly articulates a collective meaning of creativity as an insight that is found useful for some groups, which "arises from a reintegration of already existing materials or knowledge, but when it is completed it contains elements that are new" (p. 311). Stein's definition, which has been broadly accepted, captures the importance of the base of knowledge that serves as the foundation for effective creativity to take place (Runco & Jaeger, 2012). In other words, specific knowledge related to the problem at hand is indispensable to creativity. Students are able to demonstrate their creativity if and only if they master the skill in the domain. For example, an empirical study conducted in Hong Kong (Forrester & Hui, 2007) showed that Hong Kong classrooms facilitate a non-Western style of creativity. The Hong Kong experience indicated that student knowledge and intrinsic motivation are the prerequisites of creativity.

Interestingly, the personality traits leading to creativity identified by Western researchers are often found among many Asians. Those personality traits include tolerance of ambiguity, risk-taking attitude, preference for disorder, delay of gratification, and perseverance (Dacey & Lennon, 1998). Some of these traits are closely related, and they may be grouped together as follows:

1. *Tolerance of ambiguity and preference for disorder*: These are two distinct but closely related concepts. Tolerance of ambiguity means accepting the fact that a problem is inherently messy and ambiguous, such that relevant information is missing, the rules are unclear, and the solution is unavailable (MacKinnon, 1978). Different people react to an ambiguous situation differently, but creative people tend to endure and even enjoy being in this situation. Treating strangeness or vagueness as a source of excitement fosters creativity (Dacey & Lennon, 1998). In a similar vein,

prior research found that innovators often prefer complexity to simplicity (Barron & Welsh, 1952). According to Barron and Welsh, the underlying variable of this phenomenon is a preference for disorder. This does not mean that the lifestyle of creative people is chaotic. Rather, it simply means that they tend to enjoy the challenge of bringing order out of disorder. However, it is very common for teachers to provide step-to-step guidelines for students. Very often, students complain that the teacher is ineffective or the course is unstructured when he/she makes students figure out things for themselves (Weimer, 2014). Very often these students prefer straightforward guidelines to "confusing" instruction. As long as teachers assign questions similar to those discussed in class, students simply recall the modal answers to pass the exam (Bok, 2006).

But this practice is not the norm in Hong Kong and mainland China. A Chinese student who received education in both China and the US told the research team that most US teachers tended to give her very explicit instructions even though most of the time she felt she could figure it out by herself. A Hong Kong teacher interviewed by the team even used hiding information as a pedagogical strategy. Removing some relevant information does not necessarily lead to a creative solution. However, even if the solution may remain conventional, at least it can foster an attentive, problem-solving mind and the acquisition of basic, foundational knowledge, which is conducive to creativity.

2. *Risk-taking*: Playing it safe might guarantee short-term success, but in the long run, it is detrimental to creativity. Sternberg and Lubart (1995) found that risk-taking and creative performance are strongly correlated. This risk-taking attitude is well expressed by the following un-attributable quote: "If you haven't failed, you haven't tried anything new". However, some US students prefer easier classes to difficult classes (Schnoebelen, 2013). US high schools provide students with many more electives than what their Hong Kong counterparts have. Given this freedom, some US students avoid advanced math and science classes because they are afraid that those classes would drag down their GPA. To be fair, Hong Kong students are forced to take risks because the curriculum is highly standardized. However, by the time they graduate from high school, they have already built a foundation of basic knowledge and a mentality for creativity. This problem is not only found in US students. Dacey and Lennon (1998) went even further to say that very often when US teachers try to introduce innovative school projects, the superintendent or school board blocks them in order to avoid risk. On the contrary, the Hong Kong government encourages teachers to try risk-taking and innovative projects. For example, in 1998 the Hong Kong Chief Executive established the Quality Education Fund (QEF), with an allocation of HK$5 billion, to finance innovative school projects for enhancing the effectiveness of both learning and teaching. This ongoing project has inspired numerous innovations since its establishment (Hong Kong Education Bureau Central Resource Centre, 2018).

3. *Delay of gratification and perseverance*: Delay of gratification and perseverance are two different traits, but there is a close link between them. Both delay of gratification and persistence are based upon far-sighted thinking. In the classic marshmallow test (Mischel et al., 2011), young children are asked to choose between

an immediately available smaller reward (e.g., one marshmallow) and a larger reward (e.g., two marshmallows). Researchers found that children who delayed gratification for the larger reward were psychologically better adjusted, and had higher academic achievement and college scholastic aptitude test scores as adolescents. Creative people are willing to give up gratification at hand in order to achieve a higher goal in the future. Often, creative people spend a long time on a project without recognition or reward (Sternberg & Lubert, 1995). Not surprisingly, prior research affirmed that successful innovators are able to persist in spite of adversity (Csikszentmihalyi, 1996). Based on the data obtained from interviews with 31 Chinese eleventh graders, Li, Bottrell, and Armstrong (2018) found that perseverance and confidence were the common attributes of resilient Chinese adolescents, and these students indicated that their resilience was firmly grounded in Confucian cultural beliefs.

3.3 Foundational Knowledge

The authors believe that the above personality traits provide necessary, but not sufficient, conditions for facilitating creativity. A student might endure ambiguity, take difficult classes, and persist through challenging tasks, but still adopt conventional approaches; nonetheless, this kind of work ethic can enable the learner to acquire foundational knowledge for further learning and exploration. Inventor Jacob Rabinow identified three elements that are essential for creative thinking: persistence, a large knowledge base, and good judgment (Csikszentmihalyi, 1996). Besides persistence, which was discussed in the preceding section, one also needs proper domain knowledge. Otherwise one does not know what should be done with the problem at hand. Good judgment is the ability to distinguish good ideas from bad ideas. It is doubtful whether one can exercise good judgment and generate new ideas without such domain knowledge. A professor at Hong Kong Baptist University commented, "I found Chinese students more efficient and more creative. Creative students are also from France and Germany. They have *foundational knowledge*. Given advice they take initiative. They have interesting ideas on their own".

This might seem obvious to some: any learner must start with the basics. Acquiring knowledge in a given subject is the basis of expertise and creativity. Nonetheless, the perspective of domain knowledge preceding creativity has been unpopular among educators and scholars in the US. One of the popular beliefs is that one's high levels of knowledge may work against enabling expression and utilization of one's creativity (Lubart & Sternberg, 1995; Runco, 2014). Some even have assumed inevitable friction between knowledge and creativity. Some believe it is impossible for a person with abundant accumulated knowledge to break away from the already established ways of thinking or behaving (Craft, 2005; Weisberg, 1999).

In contrast to this fear of knowledge hindering creativity discussed in the US, the Chinese are familiar with the following idiom: "You start your thousand-step-journey by taking the first step". One professor who currently teaches at Hong Kong University of Science and Technology noted, "Creativity and strong technical rigor do not have to be mutually exclusive. One can learn strong technical foundation and discipline, and still develop creativity at the same time". Another professor at Hong Kong Polytechnic University, who organizes two competitions on creativity and innovation every year among Hong Kong secondary schools, provided his view on creativity and basic knowledge on the domain:

> Rigorous training and creativity are complementary. Creativity and innovation in the realm of math and science require basic knowledge. Pedagogical approach gives basic knowledge before asking students to be creative. Thus, when we teach, we teach not only basic knowledge, but also link to students' daily life. For instance, when first introducing a concept that is not visible and therefore hard to grasp, we let students do actual measurement using meters and sensors so that they get the feel of what those seemingly abstract concepts are.

Professors and teachers in Hong Kong tend to strongly agree that students need certain foundational knowledge in order to advance their learning. Compared with that of other subjects, the instruction for math is less flexible, meaning that the topics are highly structured and sequential. As a result, there are fewer pedagogical methods for math than that for other subjects (Bong, Lee, & Woo, 2015). Needless to say, this process is boring and even painful. In the US, some math teachers avoid frustrating students by relaxing the requirements of basic knowledge. But Hong Kong educators suggested the opposite. During an interview with the research team, a professor of math education at the Education University of Hong Kong said, "Drill and practice may not be the best way to learn, but it is better than nothing. Unfortunately, US students lack the basic training as what Hong Kong students have". Some Hong Kong math teachers encourage memorization of certain basic math knowledge, as shown in the following examples. By doing so, the students do not struggle with the basics and can learn faster (Cheung, 2016).

½ = 0.5 = 50%
¼ = 0.25 = 25%
¾ = 0.75 = 75%
1/8 = 0.125 = 12.5%
3/8 = 0.375 = 37.5%
5/8 = 0.625 = 62.5%
7/8 = 0.875 = 87.5%
1/16 = 0.0626 = 6.25%
25 * 4 = 100
125 * 8 = 1000

Today, with the aid of computer technology, it is unnecessary for us to store a large amount of information in our heads. Not surprisingly, many teachers tend to downplay the importance of memorization. Nonetheless, numerous studies in cognitive psychology and neuroscience indicate that working memory is related to various creative activities, and working memory training can moderate prefrontal cortex neural function for divergent thinking (e.g., De Dreu, Nijstad, Baas, Wolsink, & Roskes, 2012; Lee & Therriault, 2013; Vartanian et al., 2013). While working memory can only hold information temporarily, consistent practice of memorization can strengthen the hippocampal foundation, which is crucial for episodic and spatial memory, resulting in better neuronal plasticity. Additionally, based on PISA data, OECD (2009) found that students who employed memorization and other self-regulating strategies for studying consistently yield better learning outcomes.

3.4 Creativity in Hong Kong Science and Math Education

Traditionally, creativity has often been associated with art, music, and literature. However, researchers have suggested that science and mathematics can also promote higher creativity in students just as well as the arts (Kim, Roh, & Cho, 2016). Developmental psychologists, such as Piaget (1962) and Vygotsky (1998), have noted the importance of creativity in facilitating children's dynamic learning and cognitive development. In childhood, creative acts often happen during play; children not only enjoy the process but also take initiatives while interacting with other peers and adults. In addition, highly creative students tend to engage well with mathematics and science, and they aspire to academic success (Sarsani, 2008).

In an effort to extend creativity to the reform of math and science curriculum, starting in 2001, Hong Kong's Education and Manpower Bureau stressed the importance of creativity and critical thinking skills in schools (Curriculum Development Council, 2001). This goal was then reasserted in the 2005 curriculum guideline (Curriculum Development Council, 2005). One of the obstacles to teaching creativity in Hong Kong is its exam-oriented system, which drives students to look for the "right answer" rather than considering alternatives. As a remedy, Chan (2007) proposed flexible pedagogy and assessment systems that entail multiple-disciplinary inquiries where no fixed modal answer is given.

3.5 Combination Thinking

Although there are various approaches to cultivate creativity, the authors would like to follow the suggestion made by neuroscientists and focus on combinational thinking. According to Boden (2004), besides combinational thinking, there are two other types of creativity, namely, exploratory and transformational creativity. Exploratory thinking is thought to be the least creative type because what it entails is nothing more than generating fairly new ideas based on an existing conceptual framework. On the other hand, transformational creativity is associated with revolutionary and disruptive elements, in which totally novel ideas are generated beyond the existing conceptual space to overthrow a traditional paradigm. For example, the introduction of iTunes and the iPod by Steve Jobs changed the way people listen to music, eventually leading to the demise of certain traditional vendors, such as Tower Records. However, this type of transformational creation happens only infrequently. It is therefore unrealistic for educators to try to train every student to become someone like

Steve Jobs. Additionally, according to Zhou (2018), cognitive neuroscientists devote their efforts to analyses of associative pathways in the human brain system and yield fruitful findings related to creative behaviors. Following this direction, cognitive psychologists and neuroscientists can directly examine combinational creativity in the present, whereas exploratory and transformational creativity can only be investigated through post hoc testing. Due to the lack of relevant neuroscience data on these two types of creativity, it is difficult, if not impossible, to use them to inform educational practice or large-scale policy change.

A multi-disciplinary perspective is tied to combination thinking. In the history of invention, combination is nothing new. In this mode of reasoning, the innovator introduces new insights by combining ideas from two seemingly unrelated fields. Prior research found that people who are creative have greater associative propensities than those who are less so (Lewis, 2009; Yu, Yung, Garcia, Reyes, Bernal, & Lee, 2018). For example, telescopes were at first used by medieval military forces to monitor the movement of their enemies, and thus this tool was known as the spyglass. When no one else had explored its alternate usages, the brilliant astronomer Galileo pointed his spyglass to the sky, not the battlefield, to study heavenly objects. Then and there, he observed dark spots on the surface of the sun. This discovery paved the way to the overthrow of the Aristotelian notion of a perfect cosmos (Biagioli, 2006). A century ago, Henry Ford's invention of the assembly line revolutionized the world's auto industry. To be more specific, this innovation turned auto-making from a labor-intensive process into a mass production system. However, this idea was by no means totally original. Actually, Ford borrowed it from the butcher's conveyor belt system (Hounshell, 1984). More recently, Apple Inc.'s late founder Steve Jobs has been hailed as one of the best innovators in the world. One of his major breakthroughs was the inclusion of calligraphy into Mac's user interface. He said, "It [Mac] was the first computer with beautiful typography… If I had never dropped in on that single course [an art course] in college, the Mac would have never had multiple typefaces or proportionally spaced fonts" (Jobs, 2005). All of the above share the common thread of generating a new idea by combining two remotely associated, or even dissociated, fields: spying and astronomy, food industry and the auto industry, as well as fine art and computer science.

Contrary to popular belief, instead of exam-centric teaching, many Hong Kong teachers include learning elements that facilitate creativity by fusing seemingly unrelated disciplines together. For example, in a secondary school biology class, the teacher introduced students to biocraft model making in the following assignment, in an attempt to synthesize biology and structural engineering. The detail of the assignment is as follows (Yu, 2017):

3.6 Biocraft Model Making

3.6.1 Introduction

We have learnt about a variety of living organisms covering all five kingdoms. This project lets you demonstrate your creativity and apply your biological knowledge in making a model of hybrid animal, which can survive in a selected habitat.

3.6.2 General Instruction

1. Select a living organism that is an endangered species. Look up the scientific name of your chosen organism. Try to use Wikipedia on the Internet.
2. Search the selected habitat on the Internet to find out the environmental characteristics of this habitat. One example is given below: The organism selected is polar bear. The scientific name is Ursus Maritimus. The selected habitat is in polar regions. The environmental characteristics are winter temperatures that can reach deep into the negatives, and the winter night that can last for months.
3. Add three structural features from other organisms to change your selected organism into a hybrid organism. Explain how these structural features can help the selected organism survive in the selected habitat (i.e., increase its chance of survival). These structural features should also be shown in your model.
4. Make a fact sheet on your model of the hybrid organism and stick it onto your model.
5. The size of the model should not be smaller than 20 cm × 20 cm × 20 cm and larger than 70 cm × 70 cm × 70 cm. You should also give a ratio on the size of your model to the real hybrid organism.
6. You can use any materials to make your model. It is better to use recycled materials. Live specimens and parts from living organisms are not allowed to use.

This fun project involves a great deal of biological knowledge, and thus students had to conduct research on the characteristics of different animals, as well as how they can adapt to different environments. In other words, creative models are the result of solid knowledge. Rather than teachers telling students to memorize biological configurations and ecological attributes, students learn how to manipulate resources in order to help the animals to overcome possible hazards. Figures 3.1 and 3.2 show the products of the students.

Combination thinking can also happen across the arts and sciences. In the past, many Chinese mathematicians reexpressed mathematical theorems in the form of poetry. For example, in the Ming Dynasty, the Pythagorean theorem was illustrated in the poem shown in Fig. 3.3 (Tsui, 1997). Even though no new theorem was created by combining math and poetry in this way, this combination still makes math more

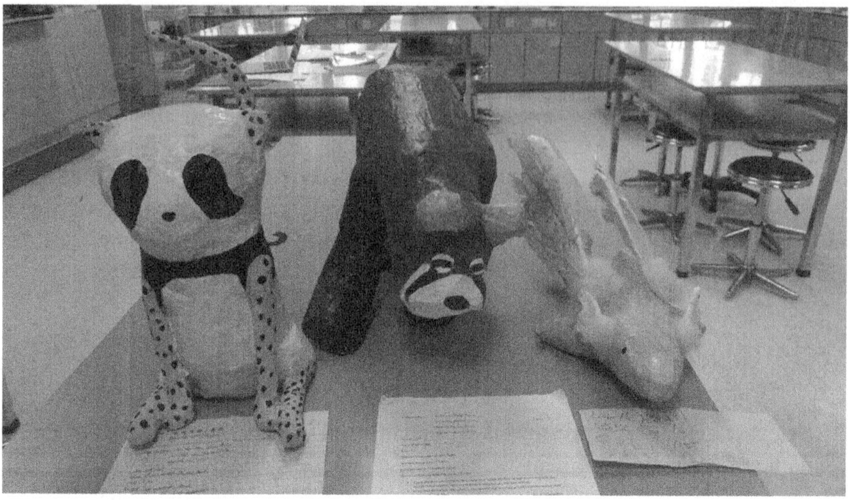

Fig. 3.1 Model of hybrid animals created by Hong Kong students: Flying fish, panda-bear, and panda-panther

Fig. 3.2 Model of hybrid animals created by Hong Kong students: Flying dog and panda-panther

今有方池一所，每邊丈二無零。 中心葭長一根肥，出水過於二尺。 斜引葭梢至岸，通然與岸方齊。 請君明算更能推，葭長水深各幾？	平平湖水清可鑒，荷花半尺出水面。 忽來一陣狂風急，吹到荷花水中偃。 湖面之上不復見，入秋漁翁始發現。 殘花離根二尺遠，試問水深尺若干？

Fig. 3.3 The Pythagorean Theorem expressed in the poetry format

interesting and accessible. By doing so, the new notion that math could be artistic and fun had arisen!

In line with this tradition, in 2016 a math teacher in Hong Kong Methodist Secondary School guided his Grade 8 students to create a *YouTube* video that incorporates singing and math. In this video, students expressed the Side Angle Side (SAS) Postulate for proving congruent triangles in a song (https://www.youtube.com/watch?v=h4XprRVDm94). As of December 9, 2019, this video has received 768,674 views, and this approach has been praised by students and educators across the city (Cheung & Chu, 2016). Inspired by the success of SAS, another Hong Kong teacher (Lam, 2018) also reexpressed scientific ideas in terms of and songs and dramas. The following are some examples:

Chemistry: https://www.youtube.com/watch?v=pvErfdOJaCg
Chemistry: https://www.youtube.com/watch?v=pbDQEuXU1vc
Physics: https://www.youtube.com/watch?v=VtJnF3tYneg
Physics: https://www.youtube.com/watch?v=4H3lutbMOY0.

3.7 Playfulness and Risk-Taking

As mentioned before, playing it safe can hinder facilitating creativity. On some occasions, Hong Kong teachers allow students to take risks by encouraging them to play around. In another exercise assigned by a Hong Kong high school biology teacher, the students were asked to correctly match a column of terms about reproduction and a column of the definition of those terms. This exercise was conducted as a speed contest. Besides knowing the definition, the learners needed to find the most efficient way to present the result. Some groups choose to cut and stick all items in both columns on a new piece of paper, whereas the fastest group cut the items in one column only and then pasted the definition to the corresponding position on the other column (see Fig. 3.4). Some solutions are efficient, while others are not. Nonetheless, failure is an option in this exercise. In addition to learning the terms, it also stimulates students to find creative ways to organize and present information.

Another prominent characteristic of science and math teaching in Hong Kong is the prevalence of low-cost, low-tech approaches. Many teachers simply use inexpensive tools, yet the expected result is still achieved and comparable to the high-tech approach. For example, a math teacher shared with the team how sticking bamboo sticks into mud could be used as a 3D model for teaching trigonometry. The math teacher asserted that this low-tech approach might even be better than the high-tech solution, which is usually a compressed 3D model on a 2D computer screen. The teacher told us that many children have difficulties in "seeing" a 3D model on a flat screen even though they could rotate the 3D model. By contrast, a physical model, such as one made of bamboo sticks, removes this barrier.

In a similar vein, a chemistry teacher used inexpensive ball and stick models to illustrate multiple bonding and creating opportunities for exploration by students. Although many Hong Kong schools are equipped with Molymod kits, they are mainly

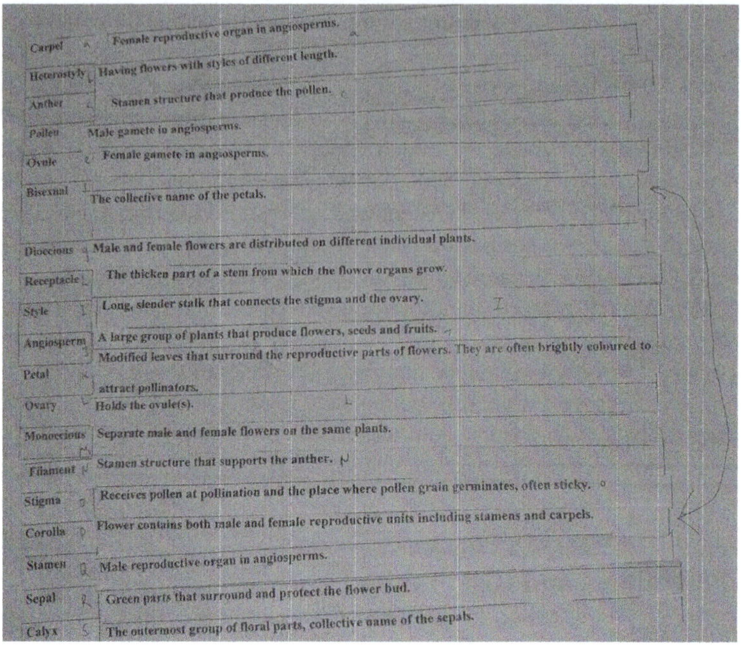

Carpel	Female reproductive organ in angiosperms.
Heterostyly	Having flowers with styles of different length.
Anther	Stamen structure that produce the pollen.
Pollen	Male gamete in angiosperms.
Ovule	Female gamete in angiosperms.
Bisexual	The collective name of the petals.
Dioecious	Male and female flowers are distributed on different individual plants.
Receptacle	The thicken part of a stem from which the flower organs grow.
Style	Long, slender stalk that connects the stigma and the ovary.
Angiosperm	A large group of plants that produce flowers, seeds and fruits.
Petal	Modified leaves that surround the reproductive parts of flowers. They are often brightly coloured to attract pollinators.
Ovary	Holds the ovule(s).
Monoecious	Separate male and female flowers on the same plants.
Filament	Stamen structure that supports the anther.
Stigma	Receives pollen at pollination and the place where pollen grain germinates, often sticky.
Corolla	Flower contains both male and female reproductive units including stamens and carpels.
Stamen	Male reproductive organ in angiosperms.
Sepal	Green parts that surround and protect the flower bud.
Calyx	The outermost group of floral parts, collective name of the sepals.

Fig. 3.4 Term-definition matching exercise that stimulates creative ways of organizing and presenting information

used for illustration only, and students rarely have the opportunity to manipulate or even construct molecular models. As a remedy, the teacher used a toy set called "varying beads" bought from Taobao (https://s.taobao.com/), as shown in Fig. 3.5, for students to manipulate 3D models in chemistry (Ng, Wong, & Mak, 2012).

Although the original intent of the toy has nothing to do with chemistry, the teacher could assign the color convention to the beads in terms of molecular models:

- Yellow or white for hydrogen
- Black for carbon
- Red for oxygen
- Blue for nitrogen
- Green for halogen

Because it is a toy set rather than a professional teaching kit, usually the sticks provided in the toy kit are too long for making molecular models. To remediate the situation, the teacher could either trim the sticks or use other sticks. Nonetheless, when both the teacher and the students "play" with the nonprofessional tool kit, it stimulates innovative solutions. Indeed, one may find similar examples in the US, such as using LEGO products for 3D modeling. The distinctiveness of this Hong Kong's teacher's approach is in utilizing self-made devices rather than adopting a standardized tool, thus further encouraging creativity.

Fig. 3.5 Beads for learning scientific models

3.8 Counterfactual Reasoning

Counterfactual reasoning is also indispensable in creativity. As the name implies, counterfactual reasoning is opposite to factual thinking, which is based on what exists. The creative aspect of counterfactual reasoning is that possible alternatives, which are contrary to what actually happened, are taken into account in order to let new ideas emerge. One famous example is that of Albert Einstein. When he was 16, he asked, "What if I could chase after a light beam?" Later, this kind of imaginative thinking paved the way for his groundbreaking research in relativity. Simply put, to stimulate creativity the learners should be encouraged to ask "what-if" questions or to "think outside the box" (Yu, Yung, Garcia, Reyes, Bernal, & Lee, 2018). Traditional researchers are opposed to counterfactual reasoning. For example, Mach (1941) said, "The universe is not twice given, with earth at rest and earth in motion; but only once, with its relative motions, alone determinate. It is, accordingly, not permitted for us to say how things would be if the earth did not rotate" (p. 284). Nonetheless, today many scholars agree that counterfactual reasoning is the key to advanced research and creative activity (Yu, 2006).

In alignment with this approach, some Hong Kong high school math teachers encouraged students to consider the alternatives by challenging conventional wisdom. For instance, in Chan's view (2015), math is bound by human convention, and mathematical definitions are constantly under revision. Thus, students should never take things for granted. For example, take the definition of the perimeter. The perimeter of a shape is the total distance around the edge of the shape. A square whose sides are 5 inches long has a perimeter of $5 \times 4 = 20$ inches. The perimeter of a circle is the circumference, which is $c = 2\pi r$. It seems that there is no other way to define the meaning of perimeter. Nonetheless, the teacher can urge learners

Fig. 3.6 A picture frame with a controversial perimeter

to think outside the box by showing a picture frame with a thick boundary and then asking them to point out the perimeter (Fig. 3.6). This is thought-provoking. Should we include the outer boundary, the inner boundary, or both as the perimeter? Can we consider different alternatives instead of fixing our minds into one single definition?

This line of reasoning can be extended to other topics in math, such as the definition of natural number, rectangle, square, and parallel line (Chan, 2015). Most mathematics textbooks do not treat zero as a natural number. In other words, the sequence of natural numbers is 1, 2, 3, 4, etc. However, what would happen if we included zero in the set of natural numbers? Additionally, many teachers differentiate between rectangle and square. By intuition, they look different, but can we treat a square as a special case of the rectangular shape (Clapham & Nicholson, 2009)?

Furthermore, when students think of parallel lines, the first picture that comes into their minds is a pair of two straight lines with a distance between them is constant along with all the points of both lines. The teacher can stimulate the learners to consider an alternative by showing the following photo (Fig. 3.7). Why must we assume that parallel lines are always straight lines? In real life, parallel curves are easily found in structures and places like railways and highways.

The authors have found this approach very refreshing. The first author of this book, who teaches university-level statistics, observed that the minds of many statistics learners are confined to linear modeling. Half a century ago, Anscombe (1973) pointed out that linear modeling may not always be true because in the real world many relationships are curvilinear in nature. When the readers are informed by Anscombe that the Pearson coefficient of x and y is 0.93 and the R^2 is 0.67, the direct interpretation is that as x goes up, y increases. However, the relationship of x and y is, in fact, curvilinear, meaning that at the inflection point y turns flat and even goes down as x increases. In spite of this warning, most statistics learners accept linearity and do not bother to consider the alternative. This narrowness could have been rectified had counterfactual reasoning been taught in high school.

It is a popular belief that mathematics is built upon invariable axioms and that therefore there is no room for open discussion. As a result, mathematics lessons are

Fig. 3.7 Parallel curves in the real world

rigid and boring, filled with jargon that seems to be irrelevant and impractical to reality. The following passage is a typical illustration of the concept "function": "A technical definition of a function is: a relation from a set of inputs to a set of possible outputs where each input is related to exactly one output. This means that if the object x is in the set of inputs (called the domain) then a function *f* will map the object x to exactly one object *f*(x) in the set of possible outputs (called the codomain)" (as cited in Kwok & Wong, 2018, p. 13). Two Hong Kong math teachers (Kwok & Wong, 2018) dared to open up the definition of function into a form of dialogue because they believed that the definitions of function have varied from time to time and from place to place, thus there is no reason that we cannot think otherwise (Siu, 1995). Their argument is philosophical: when we look around objects in everyday life, almost nothing can be clearly defined by a single definition. One may say that a typical chair consists of four legs. However, a chair might have three legs or no leg at all!

Following this line of reasoning, several innovative definitions of function arose through the dialogue. For example, the literal meaning of function is utility or effect. In this sense, *sin*(x) is the effect of *Sine* to x. In the Chinese language, "function" means containing, and therefore a function can be conceptualized as a container of numbers. When we picture a functional relationship as an equation with two sides, we can also view a function as a map or a mapping form (Kwok & Wong, 2018). Furthermore, in statistics, the standard definition of mode is the most frequently occurring value found in a set of numbers. Interestingly, Wong (2018) urged learners to consider an alternate definition. To be more specific, if the set contains "1, 1, 1,

1, 1" in a uniform distribution, what is the mode? Is there a mode at all in this case? Can the conventional meaning of mode be applied into all situations? Similarly, the teacher could urge the students to challenge other types of conventional wisdom. For example, what is a^0? What is 0!? Moreover, at first glance, the definition of height should be indisputable. Take Fig. 3.8 as an example. By the conventional definition, the vertical line indicates the length of the cone as the height.

However, when the cone is laid down (see Fig. 3.9), can we still refer to the vertical dimension as height? Also, we may ask if there is any precise definition of height.

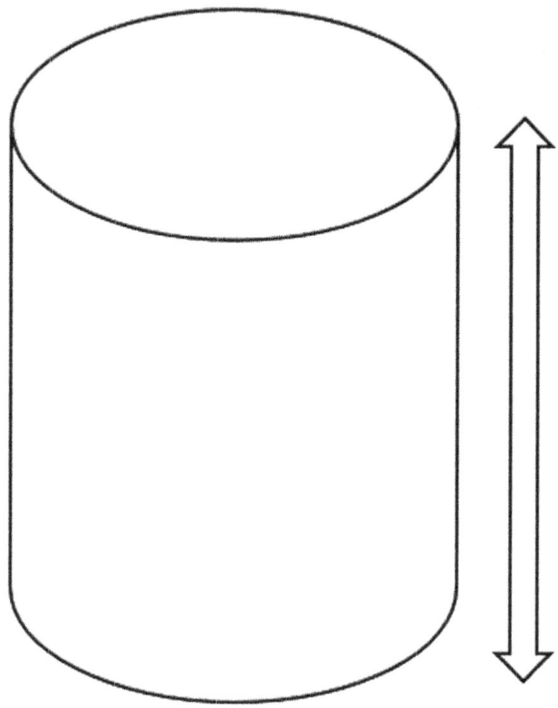

Fig. 3.8 The "height" of a cone indicated by a vertical arrow

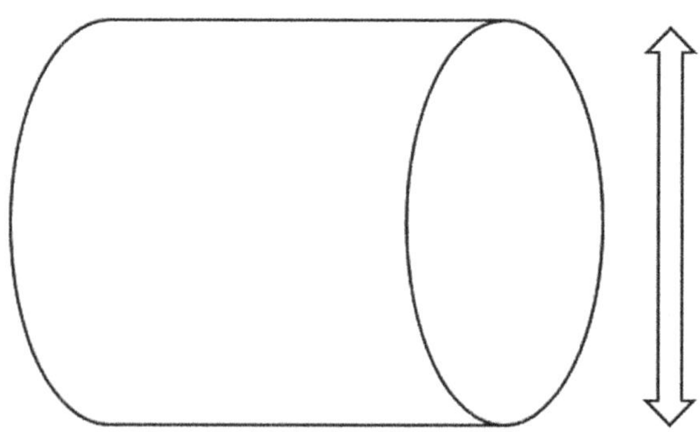

Fig. 3.9 The "height" of a horizontal cone indicated by a vertical arrow

The questions will continue to pop up when we switch up the common perspective and willing to see the other side of the same coin. This type of definition-exploration could facilitate creativity and turn a boring lesson into a lively one.

3.9 Conclusion

Professor Andrew Cohen, Director of the Institute for Advanced Study at the Hong Kong University of Science and Technology said, "Hong Kong is now perfectly positioned to attract creative researchers from around the world amid the rise of China, while America and Europe are less focused on such activities" (as cited in Chiu, 2017, para. 1). This statement may be an exaggeration, but it holds some truth, especially when some US educators move in the direction of standardization and conformity, as Puccio (2017) warned. There is a stereotype that most Asians are just "copy cats" who are not good at brainstorming, innovating, and discussing. However, as mentioned in Chap. 1, a large number of graduate students in STEM—including scientists, engineers, mathematicians, and computer programmers—are Asian Americans, including Hong Kong Chinese people. They make substantive contributions to the vibrant high-tech industry of the US. It is noteworthy that those foreign workers or immigrants received their foundational education in Asia, including Hong Kong. Contrary to popular belief, foundational knowledge and hard work are indispensable for generating creative ideas. Hong Kong educators have been trying to reposition instructional strategies by infusing combination thinking, counterfactual reasoning, playfulness, risk-taking, and other elements that lead to creativity in math and science curriculum. Are these activities, as Professor Andrew Cohen pointed out, neglected by the West, and most specifically the US? Puccio (2017) viewed conformity and standardization as the "perfect storm", but he also wrote, "While the potential for a perfect storm exists, the current overt demand for creativity creates an opportunity for educational, organizational, and societal leaders to leverage the robust body of knowledge found in the field of creativity studies. Some educational leaders and institutions of higher education are beginning to seize this opportunity" (p. 334). The authors believe that both the East and the West can benefit from a two-way exchange when it comes to creativity facilitating education.

References

Anscombe, F. J. (1973). Graphs in statistical analysis. *American Statistician, 27*(1), 17–21. https://doi.org/10.1080/00031305.1973.10478966.

Barron, F., & Welsh, G. S. (1952). Artistic perception as a factor in personality style: Its measurement by a figure preference test. *Journal of Psychology, 33,* 199–203.

Biagioli, M. (2006). *Galileo's instruments of credit: Telescopes, images, secrecy.* Chicago, IL: University of Chicago Press.

Boden, M. A. (2004). *The creative mind: Myths and mechanisms* (2nd ed.). New York, NY: Routledge.

Bong, M., Lee, S. K., & Woo, Y. (2015). The roles of interest and self-efficacy in the decision to pursue mathematics and science. In K. A. Renninger, M. Nieswandt, & S. Hidi (Eds.), *Interest in mathematics and science learning* (pp. 33–48). Washington, DC: American Educational Research Association. http://dx.doi.org/10.3102/978-0-935302-42-4_2.

Bok, D. (2006). *Our underachieving colleges: A Candid look at how much students learn and why they should be learning more*. Princeton, NJ: Princeton University Press.

Chan, D. W. (2007). Creative teaching in Hong Kong schools: Constraints and challenges. *Educational Research Journal, 22*(1), 1–12.

Chan, Y. C. (2015). Definitions and classifications of mathematical terms. *School Mathematics Newsletter, 19,* 44–58.

Cheung, M. Y. (2016). Developing the arithmetic skill of primary school students. *School Mathematics Newsletter, 20,* 64–68.

Cheung, Q. P., & Chu, S. H. (2016). The interesting aspect of math: Singing math. *Hong Kong Science Teacher Journal, 32,* 58–63.

Chiu, P. (2017, May 16). 'Time is now' for Hong Kong to promote creative research that can lead to great discoveries. *South China Morning Post*. Retrieved from https://www.scmp.com/news/hong-kong/education-community/article/2094567/time-now-hong-kong-promote-creative-research-can.

Clapham, C., & Nicholson, J. (2009). *The concise Oxford dictionary of mathematics* (4th ed.). Oxford: Oxford University Press.

Council of the Great City Schools. (2015, October 24). *Student assessments in public schools not strategic, often redundant*. Retrieved from https://www.cgcs.org/cms/lib/DC00001581/Centricity/Domain/4/Testing%20Report.pdf.

Craft, A. (2005). *Creativity in schools: Tensions and dilemmas*. New York, NY: Routledge.

Csikszentmihalyi, M. (1996). *Creativity: Flow and the psychology of discovery and inventions*. New York, NY: Harper Collins.

Curriculum Development Council. (2001). *Learning to learn—The way forward in curriculum development*. Hong Kong: Author. Retrieved from https://www.edb.gov.hk/mobile/en/curriculum-development/cs-curriculum-doc-report/wf-in-cur/index.html.

Curriculum Development Council. (2005). *The new academic structure for senior secondary education and higher education: Action plan for investing in the future of Hong Kong*. Hong Kong: Author.

Dacey, J. S., & Lennon, K. M. (1998). *Understanding creativity: The interplay of biological, psychological, and social factors*. San Francisco, CA: Jossey-Bass Publishers.

De Dreu, C. K. W., Nijstad, B. A., Baas, M., Wolsink, I., & Roskes, M. (2012). Working memory benefits creative Insight, musical improvisation, and original ideation through maintained task-focused attention. *Personality and Social Psychology Bulletin, 38,* 656–669. https://doi.org/10.1177/0146167211435795.

Edelson, P. J. (1999). Creativity and adult education. *New Directions for Adult and Continuing Education, 81,* 3–13.

Egan, K. (2005). *An imaginative approach to teaching*. San Francisco, CA: Jossey-Bass.

Egan, K., & Nadaner, D. (Eds.). (1988). *Imagination and education*. New York, NY: Teachers College Press.

Fasko, D. (2001). Education and creativity. *Creativity Research Journal, 13,* 317–327.

Forrester, V., & Hui, A. (2007). Creativity in the Hong Kong classroom: What is the contextual practice? *Thinking Skills and Creativity, 2*(1), 30–38.

Gardner, H. (1991). *The unschooled mind: How children think and how schools should teach*. New York, NY: Basic Books.

Hong Kong Education Bureau Central Resource Centre. (2018). *Quality Education Fund Cyber Resource Centre: Recent completed projects*. Retrieved from https://qcrc.qef.org.hk/project.php.

Hounshell, D.A. (1984). *From the American system to mass production, 1800–1932: The Development of manufacturing technology in the United States*. Baltimore, ML: Johns Hopkins University Press.

Jobs, S. (2005). 2005 Stanford University's commencement speech. Retrieved from https://news.stanford.edu/news/2005/june15/jobs-061505.html.

Katz-Buonincontro, J. (2012). Creativity at the crossroads: Pragmatic versus humanist claims in education reform speeches. *Creativity Research Journal, 24,* 257–265.

Kaufman, J. C., & Baer, J. (2006). *Creativity and reason in cognitive development.* New York, NY: Cambridge University Press.

Kim, M. K., Roh, I. S., & Cho, M. K. (2016). Creativity of gifted students in an integrated math-science instruction. *Thinking Skills and Creativity, 19,* 38–48. https://doi.org/10.1016/j.tsc.2015.07.004.

Kluger J. (2013, May 19). Assessing the creative spark: What Americans think about creativity. *Time, 181*(19), 58. Retrieved from http://business.time.com/2013/05/09/assessing-the-creative-spark/.

Kwok, K. L., & Wong, I. Y. (2018). Algebraic, polynomial, equations, equations, identities, and functions: Towards a dialogue with learning teachers. *Hong Kong Science Teachers' Journal, 33,* 1–16.

Lam, P. K. (2018). The theory and practice of science education songs. *Hong Kong Science Teachers' Journal, 33,* 123–141.

Lee, C. S., & Therriault, D. J. (2013). The cognitive underpinnings of creative thought: A latent variable analysis exploring the roles of intelligence and working memory in three creative thinking processes. *Intelligence, 41,* 306–320. https://doi.org/10.1016/j.intell.2013.04.008.

Lepper, M., Greene, D., & Nisbert, R. (1983). Undermining children's intrinsic interest with extrinsic rewards: A test of over justification hypothesis. *Journal of Personality and Social Psychology, 28,* 129–137.

Lewis, T. (2009). Creativity in technology education: providing children with glimpses of their inventive potential. *International Journal of Technology and Design Education, 19,* 255–268.

Li, H., Bottrell, D., & Armstrong, D. (2018). Understanding the pathways to resilience: Voices from Chinese adolescents. *Young, 26*(2), 126–144. https://doi.org/10.1177/1103308817711532.

Lubart, T. I., & Sternberg, R. J. (1995). An investment approach to creativity: Theory and data. In S. M. Smith, T. B. Ward, & R. A. Finke (Eds.), *The creative cognition approach* (pp. 271–302). Cambridge, MA: The MIT Press.

MacKinnon, D. (1978). *In search of human effectiveness: Identifying and developing creativity.* Buffalo, NY: Creative Education Foundation Press.

Mischel, W., Ayduk, O., Berman, M. G., Casey, B. J., Gotlib, I. H., Jonides, J., . . . Shoda, Y. (2011). 'Willpower' over the life span: Decomposing self-regulation. *Social, Cognitive, and Affective Neuroscience, 6*(2), 252–256. http://dx.doi.org.contentproxy.phoenix.edu/10.1093/scan/nsq081.

Ng, P. H., Wong, S. L., & Mak, S. Y. (2012). Constructing molecular models with low-cost toy beads. *School Science Review, 94,* 51–58.

Organization for Economic and Cooperation Development [OECD]. (2009). *PISA 2009 results: Learning to Learn: Student engagement, strategies, and practices.* Paris: Author. Retrieved from https://www.oecd.org/pisa/pisaproducts/48852630.pdf.

Organization for Economic and Cooperation Development [OECD]. (2011). *Lessons from PISA for the United States, strong Performers and successful reformers in education.* Paris: Author. https://doi.org/10.1787/9789264096660-en.

Piaget, J. (1962). *Play, dreams and imitation in childhood* (C. Gattegno & F.M. Hodgson, Trans.). New York, NY: W.W. Norton.

Puccio, G. (2017). From the dawn of humanity to the 21st Century: Creativity as an enduring survival skill. *Journal of Creative Behavior, 51*(4), 330–334.

Robinson, K. (2006, February). *Do schools kill creativity?* [Video file]. Retrieved from https://www.ted.com/talks/ken_robinson_says_schools_kill_creativity.

Robinson, K. (2011). *Out of our minds: Learning to be creative.* Chichester, West Sussex, UK: Capstone Publishing Ltd.

Robinson, K. (2013, April). *How to escape education's Death Valley.* [Video file]. Retrieved from https://www.ted.com/talks/ken_robinson_how_to_escape_education_s_death_valley.

Robinson, K., & Aronica, L. (2015). *Creative schools: The grassroots revolution that's transforming education*. New York, NY: Viking.
Runco, M. A. (2004). Creativity. *Annual Review of Psychology, 55,* 657–687.
Runco, M. A. (2014). *Creativity: Theories and themes: Research, development, and practice*. New York, NY: Elsevier.
Runco, M. A., & Jaeger, G. J. (2012). The standard definition of creativity. *Creativity Research Journal, 24*(1), 92–96.
Sarsani, M. (2008). Do high and low creative children differ in their cognition and motivation? *Creativity Research Journal, 20,* 155–170. https://doi.org/10.1080/10400410802059861.
Schnoebelen, A. (2013, April 26). Students avoid 'difficult' online courses, study finds. *Wired Campus or The Chronicle of Higher Education?* Retrieved from https://www.chronicle.com/blogs/wiredcampus/students-avoid-difficult-online-courses-study-finds/43603.
Siu, M. K. (1995). Concept of function: Its history and teaching. In T. F. Swetz, J. Fauvel, O. Bekken, B. Johansson & V. Katz (Eds.), *Learn from the Masters: Proceedings of Workshop on History of Mathematics Kristiansand* (pp. 105–121). Washington DC: Mathematics Association of America.
Soh, K. (2019). *PISA and PIRLS: The effects of culture and school environment*. Singapore: World Scientific.
Stein, M. I. (1953). Creativity and culture. *The journal of psychology, 36,* 311–322.
Sternberg, R. J. (Ed.). (1999). *Handbook of creativity*. Cambridge, UK: Cambridge University Press.
Sternberg, R., & Lubert, T. (1995). *Defying the crowd: Cultivating creativity in a culture of conformity*. New York, NY: Free Press.
Torrance, E. P. (1988). The nature of creativity as manifest in its testing. In R. J. Sternberg (Ed.), *The nature of creativity* (pp. 43–73). New York, NY: Cambridge University Press.
Treffinger, D. J. (1996). *Creativity, creative thinking, and critical thinking: In search of definitions*. Sarasota, FL: Center for Creative Learning.
Tsui, P. F. (1997). *Explaining poems in math*. Taipei, ROC: Ming Man Books.
Vartanian, O., Jobidon, M.-E., Bouak, F., Nakashima, A., Smith, I., Lam, Q., et al. (2013). Working memory training is associated with lower prefrontal cortex activation in a divergent thinking task. *Neuroscience, 236,* 186–194. https://doi.org/10.1016/j.neuroscience.2012.12.060.
Vygotsky, L. S. (1998). *The collected works of L.S. Vygotsky, vol. 5* (R.W. Rieber, ed.). New York, NY: Plenum Press.
Weimer, M. (2014, September 10). *She didn't teach. We had to learn it ourselves*. Retrieved from https://www.facultyfocus.com/articles/teaching-professor-blog/didnt-teach-learn/.
Weisberg, R. W. (1999). Creativity and knowledge: A challenge to theories. In R. J. Sternberg (Ed.), *Handbook of creativity* (pp. 226–250). New York, NY: Cambridge University Press.
Weisberg, R. W. (2006). Expertise and reason in creative thinking: Evidence from case studies and the laboratory. In J. C. Kaufman & J. Baer (Eds.), *Creativity and reason in cognitive development* (pp. 7–42). New York, NY: Cambridge University Press. http://dx.doi.org/10.1017/CBO9780511606915.003.
Wong, I. Y. (2018). Revisit definitions. *Hong Kong Science Teachers' Journal, 34,* 1–3.
Yu, C. (2017). G. T. (Ellen Yeung) College: 2016–17 G7 biocraft model making [Unpublished science project assignment].
Yu, C. H. (2006). *Philosophical foundations of quantitative research methodology*. Lanham, MD: University Press of America.
Yu, C. H., Yung, A., Garcia, L., Reyes, M., Bernal, A., & Lee, H. (2018, April). *Can rigorous training in math and science enhance creativity? A mixed-method approach*. Paper presented at Western Psychological Association, Portland, OR.
Zhou, K. (2018). What cognitive neuroscience tells us about creativity education?: A literature review. *Global Education Review, 5*(1), 20–34.

Chapter 4
Problem-Solving

Implementing problem-solving is one of the core objectives of Hong Kong's Learning to Learn project. Hong Kong recognizes the fundamental importance of problem-solving in math and science education. For example, every year Hong Kong's Federation of Education Workers organizes the Mathematics Creative Problem Solving Competition, aimed at providing good students with training in mathematical problem-solving skills as well as skills in collaboration, critical thinking, creative thinking, and communication (Hong Kong Education Bureau, 2018). Critical thinking is a way to enhance democracy (Santos, 2017). As mentioned in Chap. 2, it is tied to the idea of nurturing scientific citizens, who will be capable of applying critical thinking and good judgment to make informed decisions. Likely, the US has taken a similar stance as Hong Kong. The National Council of Teachers of Mathematics (NCTM) recognizes problem-solving as "not only a goal of learning mathematics, but also a major means of doing so" (NCTM, 2000, p. 52). In fact, a few decades ago, partly in response to the Mathematics Report Card (Dossey, Mullis, Lindquist, Chambers, 1988), NCTM (1989) proposed standards directing a shift in approach to mathematics education from just memorizing procedures to seeking solutions, from just memorizing formulas to exploring patterns, and from just doing exercises to formulating conjectures. In the early 1980s, a widespread awareness emerged around the importance of active incorporation of problem-solving in US math and science education. Since then, problem solving has attracted a considerable amount of national attention in the United States. In this chapter, the authors discuss how Hong Kong education and US education may differ in terms of educating their students on their problem-solving abilities, how these ideas originated from the international community (e.g., OECD), and how Hong Kong successfully put these ideas into practice.

C. H. Yu and H. S. Lee, *Creating Change to Improve Science and Mathematics Education*, https://doi.org/10.1007/978-981-15-3156-9_4

4.1 International and the US Approaches Toward Problem-Solving

It is important to point out that Hong Kong education policy makers, as mentioned in Chap. 1, are strongly influenced by international organizations (e.g., OECD), and thus not surprisingly, the Hong Kong approach toward facilitating a problem-solving mindset is based on research conducted in the West, including the US (e.g., Csapó, & Funke, 2017, OECD, 2017). The topic of problem-solving has been thoroughly studied by scholars in the West (e.g., Fischer, Greiff & Funke, 2012; Frensch & Funke, 1995, Funke & Frensch, 2007; Klieme, 2004; Mayer, 1992; Wenke, Frensch, & Funke, 2005). Researchers had already reached a consensus that a complex problem is composed of a nonroutine challenging situation, a goal state (the problem solver would like to rectify the situation in order to accomplish certain objectives), and a solution that is not readily at hand (Mayer & Whittrock, 2006). For example, a CEO of a corporation would like to reduce the environmental footprint of the production line, with the hope of slowing down or even reversing climate change. Needless to say, the problem is so complicated that no textbook-like solution to it exists. With that said, following a set of clear guidelines to report the modal answer is not considered authentic problem-solving. This perspective of problem-solving is in alignment with the view of the US National Council of Teachers of Mathematics (NCTM). According to NCTM (2000), "Problem-solving means engaging in a task for which the solution method is not known in advance. In order to find a solution, students must draw on their knowledge, and through this process, they will often develop new mathematical understandings" (p. 52).

In the 1970s and 1980s, some researchers investigated problem-solving strategies in specific domains, such as physics, writing, and chess-playing (e.g., Anderson, Boyle, & Reiser, 1985; Chi, Feltovich, & Glaser, 1981). They followed the idea developed by Herbert Simon and his associates (Bhaskar & Simon, 1977; Chase & Simon, 1973; Simon & Newell, 1971) that there was no general problem-solving ability, and hence problems in different disciplines require domain-specific knowledge and skill sets. In reaction, Funke and Frensch (2007) questioned whether this approach would hinder researchers from developing a global theory of problem-solving. Although the OECD problem-solving framework was informed by the research tradition of Herbert Simon, it leaned on the importance of global or cross-curricular problem-solving competence (Csapó, & Funke, 2017). To be more specific, in this framework, a proficient Java programmer should be able to debug a C++ or a Python program by applying the same type of logical reasoning, even though he or she never received any formal training in C++ or Python. By the same token, a capable psychology major should work very well in marketing or personnel management in a corporation by transferring his or her knowledge of human behaviors into the job, despite not having taken any class in business. A Chinese idiom shared by many Chinese teachers echoes this problem-solving framework: "When you teach a student 'one', he should be able to deduce 'two and three' from one". According to OECD, nonroutine problem-solving skills are essential twenty-first century skills.

Thus, students are expected to learn how to "work in novel environments, face problems they have never seen and apply domain-general reasoning skills that are not tied to specific domains" (Csapó, & Funke, 2017, p. 19).

Mathematical competency is considered a crucial component of problem-solving skills. Essentially, mathematics aims to look for regularity in patterns, dimensions, quantity, uncertainty, shape, and change (Csapó, & Funke, 2017). Hence, mathematics is an abstraction of the nature of the problems that we encounter in the empirical world. Usually, at the first step of problem-solving, we represent or encode the information, whereas in the end, we propose a working model, which is an abstract representation of the phenomenon under our investigation. In the abstracting of information, we inevitably engage in some form of mathematical reasoning. Some students do not see a reason for learning trigonometry, calculus, or other mathematical fields that do not seem to be practical for their everyday life and future career. This frustration is understandable when learning mathematics is treated as an end goal. Nonetheless, as mentioned at the beginning, in NCTM's (2000) view, "solving problems is not only a goal of learning mathematics but also a major means of doing so" (p. 52). To be more explicit, NCTM includes "solve problems that arise in mathematics and in other contexts" into its standards (p. 5). This pedagogical philosophy of NCTM resembles the idea of cross-curricular problem-solving competence advocated by OECD. Interestingly, all Hong Kong teachers, students, and parents interviewed by the authors also subscribe to this view. Unlike the US, Hong Kong has no high-tech industry, and therefore the applications of advanced mathematics in professional settings are limited. Nonetheless, they believe that mathematical competence is synonymous with problem-solving competence.

US educators are aware of the importance of learning mathematical problem-solving skills. For example, Richard Rusczyk founded *the Art of Problem Solving* (AoPs), an online math learning platform with more than 300,000 members who pursue and shine in math. As a former Math Olympiad winner, he also mentors America's Math Olympiad participants. Based on years of his personal experiences with kids, Rusczyk points out that students are too often taught to follow rules to get to answers, but are never taught how to have a problem-solving mentality. This strategy is counterproductive for acquiring authentic problem-solving skills. Lacking such training, which takes time, in their precollege years, students are faced with a higher and thicker wall when they enter college: a wall that requires more than what they were *told* about exceling in math (Lloyd, 2018). He has noticed a pattern, where those who delved into extra work and regularly practiced math with difficult problem-solving activities during their high school years experienced a more seamless transition into college math in comparison to those who had undergone traditional pedagogical math training. The critical factors were treating math as a problem-solving process, rather than an instant output, and not easily giving up during that process. An additional argument is that there is a shortage of qualified math and science teachers able to instill a problem-solving mindset among students. The important role of problem-solving, as the means and also the purpose of studying math and science, has been recognized and implemented by US educators and policy makers since the 1980s (Hiebert et al, 1996; Klein, 2003). Hong Kong Education

Bureau and Hong Kong educators agree as well. Nevertheless, the performance levels revealed both at domestic and international levels leaves people with uneasy questions. The authors hope to delve into how Hong Kong understands problem-solving and has successfully incorporated it into their math and science classes.

4.2 Problem-Solving: Discernment and Grit

According to Jonassen (2010), problem-solving entails multiple qualities such as a consistent and passionate curiosity toward an unseen problem and sparkling creativity in the ways the solver approaches the problem. For some Hong Kong educators, the strategy of teaching problem-solving should go one step further. Besides curiosity and creativity, one of the major components of problem-solving is *discernment*, meaning that the problem solver must understand the nature of the problem before carrying out any action. There is a widespread joke that when a man holds a hammer, every problem is a nail. Hence, problem-solving is about identifying the question and finding the appropriate route to address the question. This idea is in alignment with this well-known but unattributable saying: "Asking a good question is more important than giving the right answer".

In fact, this is closely in line with the definition of problem-solving in the view of Westerners. In spite of various definitions attributed to problem-solving, a problem is defined with some agreement. Broadly, a problem is a state where the answer or the goal is not met or attainable by the performer in an immediate, ready-made fashion (Davis, 1973; Skinner, 1966). In similar terms, it is a matter or a concept that is unknown to the responder in any given state. The nature and extent of a problem differ according to the particular situation in which a problem has arisen. The situation may range from a short pop quiz to an accumulation of interactive frictions that entail huge impacts on one's life. With some problems, people may attain the answer in a given amount of time. With other problems, such a luxury is not affordable. One good example is finding cures for certain diseases and disorders, as humans have fought to conquer them since the dawn of civilization. Furthermore, people react differently to problems with diverse approaches. Identification of a problem is the very first step; only then does the problem exist. People may not even recognize that there *is* a problem. Sometimes, people may come to such a realization but choose not to confront the problem with any action. Henry Simon (1978) called this state "accepting a task", where a human not only faces and admits the existence of a problem but also decides to progress in terms of finding the answer (p. 272). The moment the person chooses to be in this purposeful state signals the beginning of one's problem-solving, and naturally that person becomes the problem solver. The problem solver initiates observation and investigation on the problem, in the hope of acquiring some criteria and clues to desirable answers. The discernment of the solver as noted by Hong Kong educators is an essential quality of metacognition. Metacognition is loosely defined as a person's active awareness of the problem-solving process, involving observation of the problem, planning the steps for a solution, and carrying out an evaluation on the progress toward the solution, as well as their own thoughts and feelings (Fortunato, Hecht, Tittle, & Alvarez, 1991; Jonassen, 2010). Teachers can utilize metacognition training to enable students to become aware of the nature and purpose of the problem given, as well as how their minds, attitudes, and behaviors are working together while confronting it (Wilson & Clarke, 2004; Young & Worrell, 2018).

Ngai-Ying Wong, professor of Curriculum and Instruction at the Chinese University of Hong Kong and his colleagues (Wong, Marton, Wong, & Lam, 2002) provided the following insightful approach to problem-solving in the context of math learning:

> Discernment is always a function of variation. But variation in this particular case takes the form of a contrast between what is seen as mathematics and what is not. The same features could however correspond to what is experienced as variation within mathematics. Some problems can be solved within seconds, others take a lifetime; some are computational, others are not; some problems do have one single answer, others have infinitely many solutions, etc. Would the students be exposed to such kinds of variations, they would discern the features as features of (rather than constraints on) mathematics (p. 27).

While orchestrating the process of problem-solving requires discernment, exercising the entire process of problem-solving requires the quality of persistence. Hong Kong education focuses on how students learn to endure the agony and stress of problem-solving. Heppner and Krauskopf (1987) defined human problem-solving as "a highly complex, often intermittent, goal-directed sequence of cognitive, affective, and behavioral operations for adapting to what are often stressful internal and external demands" (p. 375). The process of problem-solving runs with the commitment of a person-directed and driven to the goal. It is complicated and involves stress at multiple levels. Problem-solving is the process of finding an unknown; therefore, it is never meant to be easy.

There have been numerous studies on the significance of implementing problem-solving in math and science education. The teachers give a new or seemingly twisted question, let the students delve into it for a couple of hours, and expect them to approach it in multifaceted, creative ways. Yet, they tend to neglect the fact that problem-solving calls for the solver's persistent rigorous effort and may require more than just a few hours per week. Thus, to teach rigorous problem-solving habits to students, teachers must also first engage in rigorous preparation of contents and curricula that present strenuous mental juggling. Mathematics and science studies mandate problem-solving endeavors from those who earnestly apply themselves to understand their nature and logic.

4.3 Room for Improvement in US Math and Science Education

In the following section, the authors will discuss how the current US system falls short of promoting problem-solving skills in math and science education. It is important to note here that the authors do not intend to make any overgeneralizations of all

US schools, teachers, and students. As mentioned in Chap. 1, the US education system is highly decentralized, and thus there exists a wide spectrum of educational quality. After analyzing 2007 and 2011 TIMSS math test data, Camilli and Dossey (2019) found that the US students did well in fractions in 2007 and 2011 at Grade 4. Some individual states particularly excel; when Massachusetts and Minnesota data are compared against top-performing nations in 2007 TIMSS, Massachusetts ranks number four while Minnesota is in seventh place. Nonetheless, there is always some room for improvement in any education system. The purpose of the discussion below is not to stir up afresh an atmosphere of crisis or to engage in blame-and-shame tactics. Rather, it is the sincere hope of the authors that these statistical figures and criticisms can facilitate improvement.

American College Testing, Inc. (ACT) conducted a study on large samples of American students who took the ACT in 2005–2006, which constituted about 40 percent of all graduating seniors nationwide. The report *Rigor at risk* (ACT, 2007) demonstrated further evidence of a lack of college readiness among high school seniors in terms of their core-course performances. Their suggested remedy, in order to minimize the discrepancy in essential skills between secondary and postsecondary education and to ensure a smooth transition, was to require that US high schools add more rigor into their courses. This meant not simply requiring students to take more math and science classes; rather, it entailed setting the required academic level so high that students would be expected to work with rigor to take ownership of their studies.

However, rather than pushing students to try harder, some teachers simply encourage them to give up. Some teachers comfort their students by telling them not to worry about their poor performance in math because math is not for everyone (Boaler, 2016). In neuroscience, it is not a secret that when a learner encounters a problem and makes a mistake in math, he or she grows a synapse in the brain. When this interconnected network grows, problem-solving skills are enhanced because the learner can access multiple pathways to tackle a problem (Moser, Schroder, Heeter, Moran, & Lee, 2011). Further, when the learner has an opportunity to go back to correct the error and fix the problem, he or she acquires an open and growing mindset (Mangels, Butterfield, Lamb, Good, & Dweck, 2006). To illustrate the value of encountering challenging problems, Stanford Psychology Professor Carol Dweck suggested to parents that if children come home bragging about their perfect scores in class or on a test, parents should say: "Oh, I'm sorry; that means you were not given opportunities to learn anything" (as cited in Boaler, 2016, p. 19).

Unfortunately, Dweck's approach is not widely accepted. Some US students still perceive learning math as following a set of rules rather than problem-solving. For example, half of the seventh-grade students participating in the Fourth US National Assessment of Educational Progress reported agreement or strong agreement, respectively, with the following two statements: "There is always a rule to follow in mathematics" and "Learning mathematics is mostly memorizing," (Dossey, Mullis, Lindquist, & Chambers, 1988). In a similar vein, Frank (1988) found that many

US students expect to solve any math problem in five steps or less when the step-by-step guideline is given. However, any so-called "problem" that can be solved within five steps is not challenging enough to stimulate high-order problem-solving skills.

In the 1970s and 1980s, US researchers found that some teachers simply reduced mathematical problem-solving to algorithmic recipes. In a typical mathematic classroom, the teacher explained the materials at the board, and then students tried to follow the procedures to solve the problems on their own (Dossey, Mullis, Lindquist, & Chambers, 1988). In the twenty-first century, the situation has greatly improved, but some math teachers in the US are still not well equipped with the resources and/or knowledge level to teach students, and some did not even acquire degrees in the subject (Baumert et al., 2010; Ma, 2010). These teachers fall short of carrying out and delivering to students more advanced content that requires problem-solving skills in math and science, let alone how to actively use such skills while solving math and science problems. Additionally, data from an extensive survey conducted by the National Center for Education Statistics (NCES) revealed that in the Western Region of the US, less than 25% of secondary school math teachers are considered highly qualified, while about 25–50% in the East are highly qualified. The criteria for determining whether a teacher is highly qualified include whether the teacher has a degree in the subject matter he or she teaches, teaching certification in the subject matter being taught, or 5 years of experience in teaching the subject (Gadidov, 2017).

Further, US math educator Paul Lockhart (2009) used two humorous, hypothetical examples to mock the "cookbook" or ritualistic approach used by some math teachers. In the first, a music lesson, the teacher puts some musical notes on the whiteboard. The students copy them into a different key, but they must get the clefs and key signatures right, otherwise they will not receive any credit. In the second, an art class, the teacher shows the students a worksheet of "Paint-by-Numbers". There are numbers corresponding to pigments on a piece of paper. All that the students need to do is to paint the color on the paper by following the assigned numbers! Lockhart wrote, "Sadly, our present system of mathematics education is precisely this kind of nightmare" (p. 10). This criticism might be an overgeneralization. One is hard-pressed to imagine that the average math teacher in Massachusetts and Minnesota would teach their students in this fashion. However, the first author of this book did experience similar incidents described by Lockhart. The first author teaches statistics and research methods at both the undergraduate and graduate levels. He found that many undergraduate and even graduate students are underprepared in math. Out of curiosity, he asked some of his students how they learned math in high school. Some of them told him that the teachers handed out a step-by-step procedural guide, and what they needed to do was nothing more than plugging the numbers into the equations and formulas, even though they did not understand the concepts at all.

4.4 Openness to Problem-Solving in Hong Kong Education

There is no quick fix! Students cannot learn problem-solving without hard work. A problem is inherently challenging, frustrating, and time-consuming. Either telling students to give up or using a cookbook approach makes the matter worse. The authors believe that the solution lies in openness toward and passion for problem-solving. Hong Kong secondary school curricula are well-known for their rigor. For example, the concepts of the Poisson distribution, scalars, and vectors are covered by the Hong Kong DSE assessment framework (Hong Kong Examinations and Assessment Authority, 2016), but in the US these concepts are absent from most graduate-level statistics courses, let alone high school courses. Not surprisingly, the newly developed Common Core Standards in the US do not include those difficult mathematical concepts (Common Core State Standards Initiative, 2016).

The problem-solving approach is not an empty slogan that exists only inside the documents of Hong Kong assessment framework. Below are the examples taken from secondary school teachers in Hong Kong who used it actively in classrooms.

- "Let students deduce the solution of a new problem after a demonstration of solving a similar problem."
- "Use generic learning approach, instead of specific approach, in teaching science. Draw learners' focus on the scientific pattern, and the way of critical thinking to build up scientific mindset."
- "For learning mathematics, students are invited to understand the proof of those theorems. And I would often share how these knowledge (e.g., Pythagoras' theorem) can be of use in daily life. Students are also provided with real-life examples of how statistics can be important in social policies. All these are done to invite students to think beyond textbooks."
- "Encouraging student self-inquiry, exploration with supportive measures (e.g., Lab time, post-lessons advice, etc.…)"

An analysis of PISA 2012 data revealed that Hong Kong students had internalized the problem-solving mentality. The bootstrap forest approach, which is a data mining and machine learning method, was employed to identify the most important predictors of PISA science performance in the Hong Kong ($n = 4670$) and US samples ($n = 4978$). When the sample is very large, classical statistical procedures tend to yield false claims (i.e., any trivial effect might be misidentified as significant). In addition, usually, the result deduced from traditional methods is not replicable, especially when the conclusion is based on one-shot analysis with one single sample. To remediate the issue, the bootstrap forest approach, also known as bagging, was employed in analyzing the 2012 PISA data. Bootstrapping is data-driven, meaning that random samples are repeatedly resampled and reanalyzed from the original data set. Unlike classical statistics, no assumption of the underlying distribution or external reference is required. In this sense, "bootstrap" is a concept reminiscent of pulling yourself up by your own bootstraps. In the end, the algorithm merges these repeated analyses together to obtain a converged conclusion. The bootstrap forest is so named because each individual analysis is treated as a tree. When many trees are put together, we can see the forest (the big picture or global view). There are several merits to this approach. First, the machine learning algorithm can avoid overfitting the model into the data and making false claims. Second, the analyst does not need to worry about whether the data structure can meet certain assumptions. Last, the final model results from combining numerous analyses with different subsets of the sample, thus overcoming the problem of reliance on a single analysis with a single sample in traditional statistics (Yu, Lee, Lara, & Gan, 2018).

In addition to the test of scientific reasoning, PISA developers include several surveys to collect data that are relevant to students' learning and test performance (OECD, 2013a, b). This cluster of surveys includes the student background information survey, the educational career questionnaire, the information and communication familiarity questionnaire, the parent questionnaire, and the school survey. The student survey is composed of items regarding demographics, family background, home environment, student attitude toward problem-solving, and experience at school. The educational career questionnaire is concerned with the student's career aspirations and preparation for future careers. The information and communication familiarity questionnaire consists of items relating to the availability of technology to students and their usage of technology. The parent survey entails questions pertaining to the parental background, cost of educational service, attitudes toward their child's school, parents' involvement in school, school choice, parental support for learning in the home, their child's past academic performance, and their child's career interests. The school questionnaire is completed by the principal or designee. It contains items related to the school environment, school resources, school activities, teacher qualifications, and teacher pedagogy. In the end, 399 variables, including measures of student attitudes, work habits, career aspirations, access to technology, and family and school background, were entered into the random forest model.

Table 4.1 shows the bootstrap forest result of Hong Kong students. The first column from the left contains the variable names, while the second column shows the description of the variables. The third column shows how many times this variable is identified as important after numerous repeated analyses were performed. Although the bootstrap forest stops at 117 trees, the number of splits for a variable can exceed 117 because the same variable could recur in the tree several times. The importance of the predictors is ranked in order of their frequency in the forest. The two rightmost columns illustrate the relationship between the PISA science test score and the predictor. Take openness to problem-solving as an example. If a Hong Kong student gives the response "mostly like me" or "very much like me" to the statement "I seek explanations for things", then his or her score might scatter around 578; otherwise, his or her score might be as low as 544 (see ST94Q09 in Table 4.1). It is noticeable that on the average there is a 34-point difference between these two groups of students. Similarly, if a student is positive to the statement "I like to solve complex problems", his or her score is likely to be 581; otherwise, his or her score would be 33-point lower (see ST94Q14). In other words, students who welcome problem-solving tend to perform much better in the PISA science test. A noteworthy result is that three

Table 4.1 Important predictors of PISA science test scores yielded by bootstrap forest for Hong Kong students

PISA label	Variable	# of splits	Conditions	Mean score
ST28Q01	Number of books at home	392	0–10 books >10 books	511.227 566.798
ST57Q01	Out-of-school study time for homework or other study set by the teachers	336	<2 h ≥2 h	496.554 562.116
ST91Q06	Perceived Control: "I perform poorly at school whether or not I study for my exams."	318	Strong agree/Agree Strongly disagree/Disagree	497.600 561.471
ST94Q09	**Openness for problem-solving: "I seek explanations for things."**	314	Not at all like me/Not much like me/Somewhat like me Mostly like me/Very much like me	544.295 578.628
ST94Q06	**Openness for problem-solving: "I am quick to understand things."**	285	Not at all like me/Not much like me/Somewhat like me Mostly like me/Very much like me	544.467 578.278
IC06Q01	Time spent on accessing the Internet out-of-school during weekday	272	<30 min/More than 6 h Between 31 min & 6 h	525.381 564.029
ST70Q03	Number of class periods in science per week	268	<8 ≥8	548.523 593.658
ST94Q14	**Openness for problem-solving: "I like to solve complex problems."**	262	Not at all like me/Not much like me Somewhat like me/Mostly like me/Very much like me	548.417 581.532
IC08Q03	Time spent on using email out-of-school	258	Never or hardly ever Often	525.789 563.696
IC07Q01	Time spent on using the Internet outside of school on a typical weekend day	257	<30 min >30 min	509.211 561.815

out of the top ten predictors (ST94Q09, ST94Q06, and ST94Q14) in the Hong Kong sample are related to openness to problem-solving.

The bootstrap forest result for US students is shown in Table 4.2. Although some important variables in this sample also appear in the Hong Kong sample, such as the number of books and computers at home, the US and Hong Kong results are vastly different. Notably, none of the attitudinal variables (e.g., openness to problem-solving, perceived control over school performance) are identified in the US sample. Interestingly, for the US students, several variables related to the living environment

Table 4.2 Important predictors of PISA science test scores yielded by bootstrap forest for US students

PISA label	Variable	# of splits	Conditions	Mean score
ST28Q01	Number of books at home	577	≤25 books >26 books	458.227 527.462
ST27Q03	Number of computers at home	339	None/One or two Three or more	481.926 527.158
ST91Q06	Perceived control: "I perform poorly at school whether or not I study for my exams."	331	Strongly agree/Agree Strongly disagree/Disagree	446.075 506.169
ST57Q01	Out-of-school study time for homework or other study set by the teachers	316	<6 h ≥6 h	475.600 533.841
ST71Q01	Number of all class periods a week	297	<10 ≥10	437.470 533.321
ST11Q04	Who usually lives at home with the student? Sister(s) (including stepsisters)	297	Yes No	500.025 510.965
ST11Q06	Who usually lives at home with the student? Others (e.g., cousin)	282	Yes No	464.231 514.746
ST11Q03	Who usually lives at home with the student? Brother(s) (including stepbrothers)	278	Yes No	498.703 514.095
ST26Q15	Have a guest room at home	275	No Yes	500.508 501.388

and human relationships are top predictors of test performance, but none of these are included in the Hong Kong sample. For example, if no sister, cousin, or brother is living with the student, then the student tends to gain a higher score on the PISA science test (see ST11Q04, ST11Q06, and ST11Q03 in Table 4.2). These results could be explained by the prevalence of individualism in US culture.

The 2015 PISA data are more revealing. In terms of the average test score of collaborative problem-solving, the top five East Asian nations/regions are: Singapore (561.5), Japan (551.7), Hong Kong (540.8), South Korea (538.3), and Macao (533.8). The top five Western countries are Canada (535.4), Estonia (535.2), Finland (533.9), New Zealand (532.8), and Australia (531.4). The mean score of the top five East Asian countries/regions is 545.2 with a standard deviation of 11.2. The average score of the top five Western nations is 533.7 with a standard deviation of 1.7. It is noteworthy that the standard deviation of Asian countries/regions is much greater than that of Western nations, meaning that the East is more heterogeneous than the West in regard to performance in collaborative problem-solving. This contradicts the popular belief that Oriental societies are collective and homogenous, whereas Western societies are individualists (Soh, 2019).

After the research team narrowed down potential factors of test performance using the 2012 PISA data, we focused on variables related to problem-solving in the 2015 data. In the Hong Kong sample ($n = 5{,}359$), if the teacher offers additional help for problem-solving outside regular lessons, the average PISA math and science scores of these students are much higher than those students who receive help in other ways (see Table 4.3). It is noteworthy that due to the large sample size, traditional statistical procedures (e.g., ANOVA) are not appropriate, and thus no p values are reported here. Although these data are not available in the US sample and therefore a cross-cultural comparison is not possible, this finding implies that if a teacher is willing to walk the extra mile, it could substantively boost problem-solving competence of their students.

What are the implications of the preceding analysis? Friedman (2013) suggested certain strategies to narrow the performance gap between American and Asian students. These recommended courses of action include a strong commitment to teacher training, peer learning and constant professional development, deep involvement of parents in their children's learning, an insistence by the school's leadership on the highest standards, and a culture that prizes education and respects teachers. On one hand, the authors of this book have no intention to downplay the significance of the preceding recommendations. On the other hand, when 399 variables in 2012 PISA were evaluated simultaneously in the large-scale study, another picture emerges from the data. As indicated in the results, no variables in the school questionnaire, such as teacher training, professional development, and school leadership, emerged from

Table 4.3 2015 PISA math and science scores of Hong Kong students by different types of help in problem-solving

	Math		Science	
	Mean	SD	Mean	SD
EC019Q10NA: My teacher gives hints or offers strategies that help me to solve a task				
More likely in my regular school lessons	536.38	86.44	509.95	77.67
No difference	544.94	90.26	519.97	80.01
More likely in my additional instruction	**565.61**	81.75	**537.83**	72.32
EC019Q12NA: My teacher helps me to find ways how to solve a problem				
More likely in my regular school lessons	533.88	89.04	506.94	79.55
No difference	547.88	88.89	522.11	78.61
More likely in my additional instruction	**561.74**	81.45	**536.34**	72.53
EC019Q13NA: Once we identify why I have a certain problem, my teacher provides me with a working strategy				
More likely in my regular school lessons	534.56	88.45	508.63	77.75
No difference	546.55	89.74	520.41	80.11
More likely in my additional instruction	**563.18**	80.51	**537.21**	72.13

the bootstrap forest. Simply put, they paled in comparison to student-level variables, home environment variables, and technological factors, all of which are outside of school. The authors by no means seek to dissuade government agencies from investing resources in schools or improving teacher qualifications. Neither do the authors assert that teachers and school administrators have no role in helping students achieve in their science learning. Nonetheless, given the greater importance of other variables, perhaps attention to these factors may yield effective outcomes. For the last several decades, US schools have been investing tremendous resources in instructional technologies, such as hypertext, multimedia, and mobile computing (e.g., iPad), but US student performance on both PISA and TIMSS stalled. 2015 PISA data further reveal that teachers could play a significant role in improving problem-solving abilities of students when additional help is provided. Based on these salient findings, teachers, school administrators, and parents can work hand in hand to create a home and school culture that promotes openness to problem-solving. Next, concrete examples from Hong Kong will be illustrated.

4.5 Examples of Effective Hong Kong Math and Science Education

4.5.1 Arbitrary Math for Promoting Critical Thinking

To a certain extent, mathematics is arbitrary; nonetheless, for some Hong Kong math teachers, this arbitrariness could yield teachable moments for encouraging problem-solving and critical thinking. Fung (2014) used the following fun examples to demonstrate the arbitrariness of math.

Case 1:
Student: How many sides does a square have?
Teacher: Four
Student: Why?

Case 2:
Student: Why is 0 an even number?
Teacher: Because when 0 is divided by 2, there is no remainder.

Case 3:
Student: What do the signs a|b mean?
Teacher: It means a is divided by b.

Case 4:
Teacher: A polite number is a positive integer that can be expressed as the sum of two or more consecutive positive integers.
Student: Why is it called "polite"?

The symbols and definitions in the preceding examples belong to a particular culture or community of mathematicians. Although these ideas and technical terms had been commonly accepted by mathematicians, nonmathematicians become confused by the jargon and grow skeptical of these arbitrary conventions (Hewitt, 2002). There are different approaches to face this challenge. One of them is simply to urge students to accept the teacher's authority and live with the confusion, or the teacher could turn the tables by using the confusing and arbitrary scenarios to promote critical thinking (Fung, 2014):

Case 1: Ask students to explain why there are four sides in a square by listing the conditions of a square. Students can be hinted to use a parallelogram and the "if and only if" (iff) condition for proving that there can only be four sides in a square.

Case 2: The teacher can ask the students to define the target of the study (a|b where a and b are both integers). Next, the learners are urged to use "iff" to define the characteristics of the mathematical objects (e.g., $a \neq b$ and b can be a without a remainder).

Case 3: The teacher can give a set of even numbers (6, 8, 24, 2, 0…etc.) and a set of non-even numbers (3, 7, 1, 1.4…etc.) to the students. Next, the teacher asks the student to divide each of the given even numbers by 2. Afterward, the teacher asks the learners to develop a set of questions to investigate the characteristics of even numbers. The following are some sample questions asked by students.

- Is it true that the sum of two even numbers must be even?
- Is it true that any product of two even numbers must be even?
- Now there is an even number. We know that this even number is the sum of two integers. Is it necessarily true that these two integers are not even numbers?
- Must the difference between two even numbers (the big one minus the small one) be an even number?
- If two non-even numbers are added together, would the result be always uneven?
- Now there is a whole number that is not even, we know that this whole number is the sum of two integers. Is it necessarily true that these two integers are not even numbers?

Case 4: The teacher can ask students to review the definition of polite numbers, and then ask them to answer the following questions:

- Is 9 a polite number?
- Is 30 a polite number?
- Is there any impolite number? If so, please explain it by example.
- Is there any way to determine whether a number is polite or not.

Just like any human convention, mathematics is full of seemingly arbitrary and unexplainable concepts and symbols, which might hinder students from further learning. When teachers appeal to authority or tradition, this might even lead to resentment and rejection. Conversely, some Hong Kong teachers have used these humorous cases as a starting point for learning critical thinking and problem-solving.

4.5.2 *Trick and Treat*

As mentioned previously, one of the common problems in math learning is that many students blindly follow the rules without thinking through the problem or evaluating the situation. This phenomenon connects to the lack of discernment utilized by students. Instead of being vigilant problem solvers who suspect and inspect every aspect of a given problem, students allow themselves to be assembly workers conditioned by familiar-looking products, ready-made formulas, and unattended pathways of thinking. Once a Hong Kong teacher tricked his students by asking this silly question: "There are 75 cows and 34 goats on the ship. What is the age of the Captain?" Many students immediately answered: "75–34 = 41". The lesson of this story is that teaching rules, formulas, and equations in a mechanical fashion can turn students into mindless robots that are incapable of critical thinking. Very often the first author of this book encounters students copying and pasting numbers from the computer output without careful thinking, even though the output is nonsensical. For example, many statistics students report the probability as .000 because this is what SPSS shows, but in fact, the *p* value can never be zero because the two tails in any sampling distribution never touch down the x-axis given infinite repeated samplings.

Inspired by a trick question used in the Microsoft screening exam, a Hong Kong educator created a teachable moment for problem-solving (Cheung, 2017). The trick problem used by Microsoft is as follows: Given Fig. 4.1, what is the area of the triangle?

Many interviewees for Microsoft positions applied this formula: (h * b)/2 to answer the question. However, this is the wrong answer because in the first place this triangle cannot exist at all! If the length of the diagonal line is 12 cm, at most the orthogonal line to the diagonal line can be as long as 6 cm only. Put it bluntly,

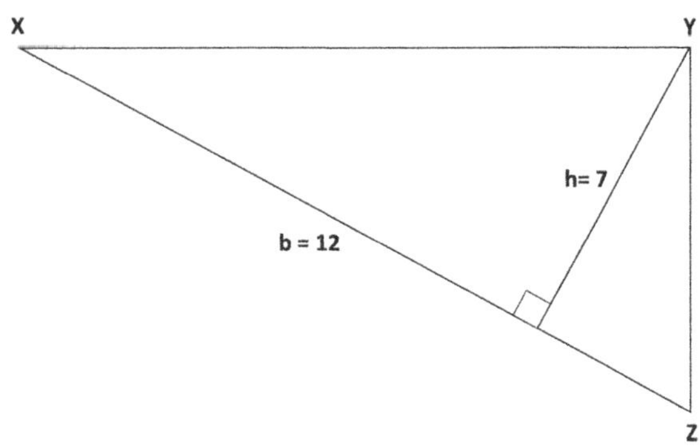

Fig. 4.1 An impossible triangle for tricking students

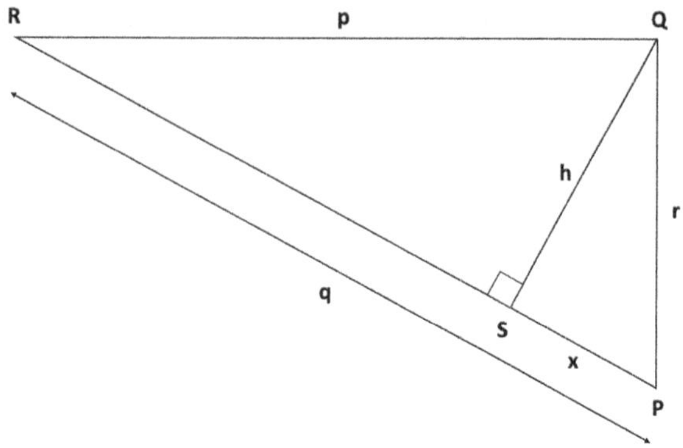

Fig. 4.2 A triangle for problem-solving (Adapted from Cheung, 2017)

those interviewees failed to evaluate the situation critically and jumped into a hasty conclusion.

Cheung (2017) found that although students who were tricked might be shocked by their carelessness, this could become a valuable teachable moment for problem-solving and critical thinking. The teacher can discuss with the students why the length of that line could not exceed 6 cm. There are multiple solutions to this problem. One of the solutions is illustrated in the following (see Fig. 4.2):

$$h/(q - x) = x/h$$
$$(q + 2h)(q - 2h) \geq 0$$
$$q + 2h \geq 0$$
$$q - 2h \geq 0$$
$$1/2q \geq h$$

"Trick questions" are regarded as something negative by many students. However, to help students learn problem-solving and critical thinking, the teacher must go beyond any conventional, formulaic approach. Once in a while, the learners need to be tricked so that they can realize their habitual thought patterns and acquire advanced problem-solving skills.

4.5.3 *Paradox and Unsolvable Problems*

Many American students equate math with a set of axioms, formulas, and equations that can deduce the correct answer in an absolute sense, but actually, math is more about concepts and logical reasoning. Students who get used to certainty will find advanced math and science difficult because advanced knowledge entails paradoxes and unsolvable problems. Hence, one of the essential attitudes for problem-solving is to admit that not all problems can be solved. For example, according to Gödel's theorem, a system is either complete or consistent (Chaitin, 1998). In other words, no formal system of mathematics can be internally consistent and contain all possible mathematical truths at the same time, because a system cannot prove its own truths with reference to its own theorems. In our experience, we intuit objects as complete and whole but indeed only parts of an object are reachable to us. In order to learn Gödel's theorem, the learner must understand the problem of self-reference. Interestingly, a Hong Kong teacher is not afraid of confusing his students by including self-referential problems in his math lessons and assessments (Leung, 2015). Self-reference is a property of a statement where it refers to itself or its own referent. A typical example is: "This sentence is not true". Needless to say, it is very difficult to judge whether this sentence is true or not. A variant of this self-referential paradox is: "I am lying". Again, it is virtually impossible to tell whether the statement is true or false. If this statement is true, then I am really lying. But if I am a liar, then this sentence is a lie and thus it should be false. Similarly, Leung (2015) created the following trick questions:

Which of the following is true? Which one is false?

1. Only Statement 1 is false.
2. Only Statement 2 is false.
3. Only Statement 3 is false.
4. Only Statement 4 is false.
5. All five statements are false.

Which of the following is true? Which one is false?

1. Only Statement 1 is false.
2. Only Statement 2 is false.
3. Only Statement 3 is false...

(n-1). Only (n-1) is false.
(n). All the statements are false.

Undoubtedly, this type of question is unconventional in high school math lessons, because there is no solution to these paradoxes. Nonetheless, these kinds of trick questions could promote high-order problem-solving skills.

Leung is not alone. Other Hong Kong high school teachers also utilized paradoxes, such as Buffon's Needle Problem (Buffon, 1777) and Bertrand's Paradox (1889), to convey the message that some math problems are ill-posed and thus keeping an open mind is advisable (Lee, 2015). In Buffon's needle problem, the problem solver is asked to find the probability that a needle of length x will land on a line, given a floor with parallel lines equally spaced at distance y apart. There is a solution if the sampling space or the population is finite, X = {x1, x2, x3...xn}. Given a finite sampling space, we can easily employ the principle of indifference to assert that every element in the set has an equal chance to be taken into the event. However, when the sampling space or the population is infinite, the problem becomes really problematic (Lee, 2015). Many American teachers adopt the simplistic approach to probability, telling students that every element of a set has an equal chance to take part in the event; thus, when there are 100 people in the population, the probability of being sampled is 1/100. This definition works well in a closed system. However, a real population is open-ended rather than fixed. For instance, if the population consists of all people in the US, then every single second some people die and other people are born. When the first author introduced alternate perspectives to probability and randomness (Yu, Anthony, & Behrens, 1995), most students were puzzled by this uncertainty and reverted to the simplistic version. This dampens problem-solving!

The Bertrand paradox is as follows: Given an equilateral triangle inscribed in a circle, if a chord of the circle is chosen at random, what is the probability that the chord is longer than a side of the triangle? There are at least three solutions to the problems, but all three answers are different: 1/2, 1/3, 1/4. Which answer is correct depends on how randomness is defined (random endpoints, random radius, or random midpoints). Without a clear definition of randomness, there would not be a definite answer (Marinoff, 1994). This pedagogical approach is mind-boggling to students used to conventional approaches. Yet it is the conviction of these authors that the Hong Kong approach can pave the way for learning advanced problem-solving skills.

4.5.4 George Polya's Problem-Solving Strategy

From our interviews, we found that at least two Hong Kong teachers adopted the problem-solving strategies suggested by George Polya. According to Polya (1957), to solve a problem one can start from requirements, assumptions, and facts, and then draw consequences from them, and more consequences from the consequences, until one reaches the starting point. Simply put, the problem solver needs to gather as much given information as possible in order to deduce from the known to the unknown. Polya (1954a, b) realized that the conjectures used for inferences might not be factual at all. Hence, one could consider the analogy of two theorems, say A and B, to discover a common ground H, from which both A and B would follow: A is implied by H, and B is implied by H. But, we do not have H; we just hope there is such an H. According to Polya, we could reach a reasonable conclusion expressed by the following pattern:

- A is implied by H
- B is implied by H
- B is true
- A is more credible

Specifically, this problem-solving strategy entails a two-step process. The first step is to conduct an end-means analysis by employing regressive reasoning, as opposed to the popular means-end analysis. When the end-means analysis is converted into action, the components of the execution plan have to be synthesized in forward order. This execution plan is called "synthesis" (Yu, DiGangi, Jannasch-Pennell, 2008).

A Hong Kong teacher (Wai, 2011) used the following example to demonstrate Polya's problem-solving approach (see Fig. 4.3):

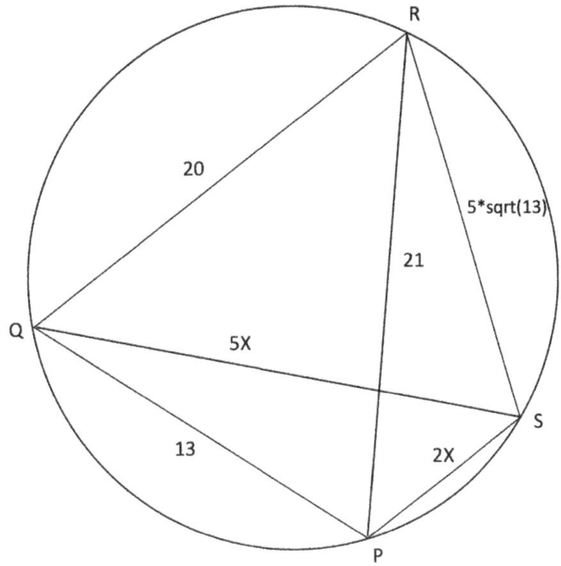

Fig. 4.3 A quadrilateral inside a circle (Adapted from Wai, 2011)

Problem: Fig. 4.3 shows a quadrilateral PQRS inside a circle.
Given: PQ = 13, QR = 20, and PR = 21.

If the lengths of PS, QS, and RS are 2x, 5x, and 5*sqrt(13), find x.

Actually, there are multiple pathways to solve this problem based on the known facts. The teacher could encourage the learners to explore alternate routes. One of the solutions is based on Ptolemy's theorem (see Fig. 4.4)

If a quadrilateral is given with four vertices *P, Q, R*, and *S* in order, then the theorem states that $(PQ)(RS) + (QR)(PS) = (PR)(QS)$

13(5*sqrt(13)) + 20(2x) = 21(5x)
65x = 65*sqrt(13)
x = sqrt(13)

If the student does not know the Ptolemy's theorem, the alternative is a combination of Sine theorem and Cosine theorem:

Sine theorem: p/Sin P = q/Sin P = r/Sin R = *diameter*
Cosine theorem: $p^2 = q^2 + r^2 – 2qr \text{ Cos } P$

During the problem-solving process, the teacher could direct students to pay attention to the known facts and boldly make conjectures in spite of the uncertainty. Some students might feel that mathematics is detached from reality and fail to see a reason to learn it. Nonetheless, the teacher could emphasize that mathematical problem-solving skills can go beyond math. According to George Polya (1954a), all human knowledge outside mathematics and logic are built upon conjectures. Chances are that these conjectures are also based on other conjectures. In contrast to demonstrative reasoning in mathematics and logic that yields certainty, many other

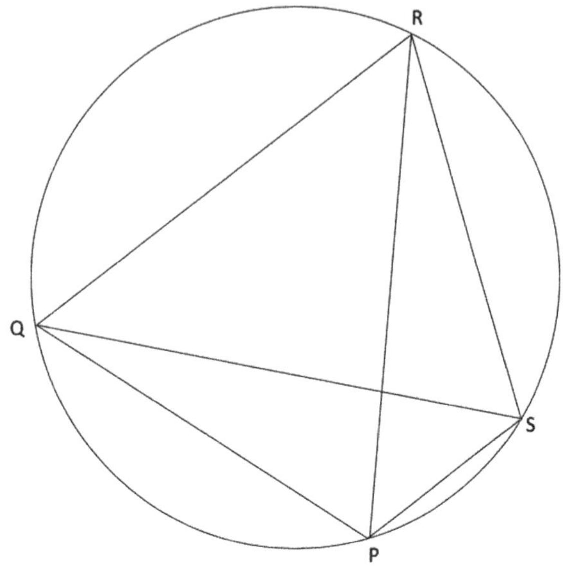

Fig. 4.4 Ptolemy's theorem concerns the relationship among the lengths of vectors in a cyclic quadrilateral

inferences belong to the realm of plausible reasoning, which is controversial and provisional, such as a lawyer's circumstantial evidence and a historian's documentary evidence. Indeed, a lawyer and a historian have to propose plausible explanations based on thin data after the fact.

4.5.5 Science Camp

Students in Hong Kong benefit from a variety of summer intensive programs and leadership camps. An example is a three-day camp that aims to cultivate students' attitude development, not just their academic knowledge and skills. On day one, the recruited trainees form a team and begin to draft a proposal after being exposed to a certain topic or issue. The project objective is to remodel the outdated mall in a city. The following day, teams share their proposals in the various forms of presentations and are allocated available and appropriate resources. With the plans, they have developed and the resources they have acquired, trainees devote the rest of the second day and the last day implementing their own solutions to the problem. During the final hours of the camp, all teams present their solution models and are evaluated by their peers and teachers. The intention of this short journey is to give trainees the lesson that problem-solving, although usually a rigorous, intensive, and time-sensitive process, can entail mutual work, interactive thinking, and the exchange of creative and well-organized ideas. One of the main objectives of this camp is to promote independent critical thinking. Nevertheless, trainees are not left to themselves. During the session, they are surrounded and mentored by instructors who are willing to guide them if requested.

In the beginning, trainees attend a workshop to identify and understand the issue and receive mandatory basic training between student-driven problem-solving activities. Teachers provide the foundational knowledge and general guidelines of the project, and students are then allowed to solve the problem independently and creatively.

4.5.6 *Chemistry Experiment*

In the integrated science class at a Hong Kong high school, a teacher interviewed by the research team gives his students a series of quite challenging tasks to provoke their problem-solving mindset. The central objective of the tasks is to conduct a fair test in scientific investigation, in which the experimenter changes one variable at a time while leaving the other conditions unchanged. Here, students are to run a simple experiment that mandates a fair test. For instance, they are to investigate the relationship between the rise in temperature of a tube of water and the area of the open-air hole of a Bunsen burner. Teachers instruct the students about the hypothesis, the variables involved (such as dependent, independent, and control variables), the general procedures of the experiment, and the precautions. The experiment begins with students and teachers wearing white coats and protective goggles. However, with a twist, the teacher creates experimental bias or mistakes in the protocol while students are observing. The table is now turned as the teacher asks students if the experiment that he or she just conducted displayed any source of errors. When the errors are detected by the students, the teacher asks how to improve the experimental design to make a fair test.

This learning by troubleshooting approach is similar to the Cisco Certified Internetwork Expert (CNIE) exam, Cisco's highest certification. The exam entails a written test and a lab test. In the lab test, the network might be purposefully misconfigured and thus dysfunctional. Then the examinee is asked to reconfigure the network to restore its function. The eight hour lab test is designed to find out whether the candidate knows inter-networking theory and demonstrates advanced hands-on mastery (Cisco Systems, 2019). It is important to point out that this approach is used for the highest level of certification for network engineers in the US, but Hong Kong teachers use it for high school students! The philosophy of this pedagogical method is in sharp contrast to the common practices in the US, where an exam guide is given to the students beforehand, and the questions on the guide are expected to be very similar to those on the exam. If even a few words are changed, the student does not know what to do, let alone conduct troubleshooting in a lab or redesign an experiment.

4.5.7 *Crime Scene Investigation*

The Interdisciplinary Forensic Science Programme used by a group of Hong Kong teachers is a good example of teaching science based upon an integrative and problem-solving approach (Grobler, 2016a). In this program, which was composed of six sessions, students were urged to employ knowledge of physics, chemistry, and biology to investigate a "murder case". A problem is inherently ill-defined and ambiguous, meaning that there is no preset or standard way to resolve the issue. Usually, this problem must be approached from multiple analytical contexts and angles of reasoning to gain a complete picture of the phenomenon (Laxman, 2013). A murder mystery is a typical challenging problem. Along this line of reasoning, the father of Exploratory Data Analysis, John Tukey (1977), equated scientific inquiry to detective work. In a similar vein, Josephson and Josephson (1996) also related abductive reasoning in inquiry to detective work. In EDA, the role of the researcher is to explore the data in as many ways as possible until a plausible "story" emerges. Detectives collect related data about suspects and circumstances based on keen powers of observation (Yu, 2017). While Tukey and Josephson and Josephson used detective work as a metaphor, Hong Kong teachers implement crime scene investigation in their pedagogical strategies.

The inquiry process began with a crime scene (a "drug factory") where a "dead body" has been found. In the first session of the investigation, which was guided by a physics teacher, students explored the use of camera lenses and microscopes to collect initial evidence. In the second session, which centered on applications of chemistry and biology, students used various chemical methods, such as dusting, superglue fuming, iodine fuming, and acidified hydrogen peroxide, to lift latent fingerprints from different surfaces like glass, paper, and metal. Next, students employed knowledge of biology to examine and match different sets of fingerprints in order to identify the suspects. In the third session, students used thin layer chromatography to compare the unknown white powder collected in the crime scene to other known samples in order to identify the drugs. The fourth session shifted gears to biology again. Students conducted blood tests to determine the blood types of different suspects and to match it to the blood type found at the crime scene. The fifth session was also concerned with biology. The focus of this investigation was Forensic Dentistry. During this process, students were asked to match the bite marks of the apples found in the crime scene to the suspects' teeth. The last session was also a biology session, in which students compared the DNA left at the crime scene with that of the suspects.

In this program, students were not passive observers or listeners; rather, they learned the scientific methods by exploring different possibilities. Further, students were empowered by being allowed to use different scientific instruments that were normally inaccessible to them, such as micropipettes, a centrifuge, and gel electrophoresis. The teachers reported that this program intensified students' interest in science because detective work is fun (Grobler, 2016b). It does not matter whether or not these students will become forensic scientists or detectives in the future. This program led them to be aware that in the real world, knowledge is not compartmentalized (physics, chemistry, biology), and the scientific method is the reasoning mode or mentality that enables us to synthesize different domains of knowledge for problem-solving.

4.6 Conclusion

When students are only exposed to a series of similar math problems, they are not allowed to expand their mathematical brainstorm. Repeated exercises may serve as a useful phase when students are first taught about the basic foundation of certain mathematical rules. However, students are then too familiarized with math problems and are conditioned to a standardized pathway of thinking that leads to a quick answer. Then math becomes nothing more to students than memorization of a set of equations. Math teachers should not make the mistake of raising students to be prototypical math-robots who are only capable of operating at pre-programmed levels. Instead, instructors should be able to train students as human troopers who are ready to fearlessly take down alien math problems. Such training should begin from an early age, to confront new, multifaceted, advanced problems with passion and grit.

By the same token, science should not be taught as a matter of following a set of experimental protocols and memorizing a collection of terms. Hong Kong educators are well aware that memorization of terms and formulas does not facilitate scientific research and problem-solving. Take President Lap-chee Tsui of the University of Hong Kong as an example. In 1989, Professor Tsui became internationally renowned for his groundbreaking research on the cystic fibrosis transmembrane conductance regulator (CFTR), the defective gene that causes cystic fibrosis. Additionally, he made substantive contributions to human genome research, such as the characterization of chromosome and identification of other disease genes. However, he was not good at memorizing jargon in biology, and as a result he earned the third honor only upon his graduation from the university. Professor Tsui attributed his accomplishments to his exploratory and experimental mentality, as well as his openness to problem-solving (Wong, 2018). Inspired by Professor Tsui, many Hong Kong teachers emphasize problem-solving skills, such as using trick questions, arbitrariness in math, and paradoxes to create teachable moments. We could be inspired by Professor Tsui too.

References

American College Testing [ACT], Inc. (2007). *Rigor at risk: Reaffirming quality in the high school core curriculum*. Iowa City, IA: Author.

Anderson, J. R., Boyle, C. B., & Reiser, B. J. (1985). Intelligent tutoring systems. *Science, 228*, 456–462.

Baumert, J., Kunter, M., Blum, W., Brunner, M., Voss, T., Jordan, A., … & Tsai, Y. M. (2010). Teachers' mathematical knowledge, cognitive activation in the classroom, and student progress. *American Educational Research Journal, 47*(1), 133–180.

Bhaskar, R., & Simon, H. (1977). Problem solving in semantically rich domains: An example from engineering thermodynamics. *Cognitive Science, 1*(2), 193–215.

Boaler, J. (2016). *Mathematical mindset: Unleashing students' potential through creative math, inspiring messages and innovative teaching*. New York, NY: Jossey-Bass.

Buffon, G. (1777). Essai d'arithmétique morale. *Histoire naturelle, générale er particulière, Supplément, 4,* 46–123.

Camillia, G., & Dossey, J. (2019). Multidimensional national profiles for TIMSS 2007 and 2011 mathematics. *Journal of Mathematical Behavior* [Online First]. https://doi.org/10.1016/j.jmathb.2019.02.001.

Chaitin, G. J. (1998). *The limits of mathematics: A course on information theory and the limits of formal reasoning*. Singapore: Springer.

Chase, W. G., & Simon, H. A. (1973). Perception in chess. *Cognitive Psychology, 4,* 55–81.

Cheung, Q. P. (2017). Break the convention: A question in the interview. *School Mathematics Newsletter, 21,* 64–76.

Chi, M. T. H., Feltovich, P. J., & Glaser, R. (1981). Categorization and representation of physics problems by experts and novices. *Cognitive Science, 5,* 121–152.

Cisco Systems. (2019). *Expert certification*. Retrieved from https://www.cisco.com/c/en/us/training-events/training-certifications/certifications/expert.html.

Common Core State Standards Initiative. (2016). *Mathematical standards*. Retrieved from http://www.corestandards.org/wp-content/uploads/Math_Standards1.pdf.

Csapó, B., & Funke, J. (Eds.). (2017). *The nature of problem solving: Using research to inspire 21st century learning, educational research and innovation*. Paris: OECD Publishing. https://doi.org/10.1787/9789264273955-en.

Davis, G. A. (1973). *Psychology of problem solving: Theory and practice*. New York, NY: Basic Books.

Dossey, J., Mullis, I., Lindquist, M., & Chambers, D. (1988). *The mathematics report card: Are we measuring up? Trends and achievements based on the 1986 national assessment*. Princeton, NJ: Educational Testing Service.

Fischer, A., Greiff, S., & Funke, J. (2012). The process of solving complex problems. *Journal of Problem Solving, 4*(1), Article 3. http://dx.doi.org/10.7771/1932-6246.1118.

Fortunato, I., Hecht, D., Tittle, C. K., & Alvarez, L. (1991). Metacognition and problem solving. *The Arithmetic Teacher, 39*(4), 38–40.

Frensch, P. A., & Funke, J. (1995). *Complex problem solving: The European perspective*. New York, NY: Erlbaum.

Friedman, T. (2013, October 22). The Shanghai secret. *New York Times*. Retrieved from http://www.nytimes.com/2013/10/23/opinion/friedman-the-shanghai-secret.html?_r=0.

Fung, C. Y. (2014). Mathematical knowledge is "autocratic". *School Mathematics Newsletter, 18,* 19–34.

Funke, J., & Frensch, P. A. (2007). Complex problem solving: The European perspective - 10 years after. In D. H. Jonassen (Ed.), *Learning to solve complex scientific problems* (pp. 25–47). New York, NY: Erlbaum.

Gadidov, B. (2017, April). *Predicting highly qualified math teachers in secondary schools in the United States*. Paper presented at SAS Global Forum, Orlando, FL.

Grobler, G. C. (2016a). Interdisciplinary Forensic Science Programme. *Hong Kong Science Teachers' Journal, 31,* 14–18.

Grobler, G. C. (2016b). Let our students teach! Teaching as a way of mastering knowledge. *Hong Kong Teacher's Journal, 32,* 79–83.

Heppner, P. P., & Krauskopf, C. J. (1987). An information-processing approach to personal problem solving. *Counseling Psychologist, 15,* 371–447.

Hewitt, D. (2002). Arbitrary and Necessary: A way of viewing the mathematics curriculum. In L. Haggarty (Ed.), *Teaching mathematics in secondary schools: A reader* (pp. 47–63). London: Routledge.

Hiebert, J., Carpenter, T. P., Fennema, E., Fuson, K., Human, P., Murray, H., … & Wearne, D. (1996). Problem solving as a basis for reform in curriculum and instruction: The case of mathematics. *Educational Researcher, 25*(4), 12–21.

Hong Kong Education Bureau. (2018). *HK Mathematics Creative Problem-Solving Competition*. Retrieved from https://www.edb.gov.hk/en/curriculum-development/major-level-of-edu/gifted/resources_and_support/competitions/local/cps.html.

Hong Kong Examinations and Assessment Authority. (2016). *Category A: Senior secondary subjects*. Retrieved from http://www.hkeaa.edu.hk/en/hkdse/assessment/subject_information/category_a_subjects/hkdse_subj.html?A1&1&4.

Jonassen, D. H. (2010). *Learning to solve problems: A handbook for designing problem-solving learning environments*. New York, NY: Routledge.

Josephson, J. R., & Josephson, S. G. (Eds.). (1996). *Abductive inference: Computation, philosophy, technology*. Cambridge, UK: Cambridge University Press.

Klein, D. (2003). A brief history of American K-12 mathematics education in the 20th Century. In J. M. Royer (Ed.), *Mathematical Cognition* (pp. 175–259). Charlotte, NC: Information Age Publishing.

Klieme, E. (2004). Assessment of cross-curricular problem-solving competencies. In J. H. Moskowitz & M. Stephens (Eds.), *Comparing learning outcomes: International assessments and education policy* (pp. 81–107). London: Routledge.

Laxman, K. (2013). Epistemology of well and ill structured problem solving. In Sebatien Helie (Ed.), *The psychology of problem solving* (pp. 3–13). New York, NY: Nova Science Publishers Inc.

Lee, C. M. (2015). Geometrical probability and Bertrand's Paradox. *School Mathematics Newsletter, 19,* 14–27.

Leung, K. S. (2015). You will never get a zero in this exam no matter what your answer is. *School Mathematics Newsletter, 19,* 5–13.

Lloyd, C. (2018, June 21). *Why our smartest students are failing math*. Retrieved from https://www.greatschools.org/gk/articles/why-americas-smartest-students-fail-math/.

Lockhart, P. (2009). *A mathematician's lament: How school cheats us out of our most fascinating and imaginative art*. New York, NY: Bellevue Literary Press.

Ma, L. (2010). *Knowing and teaching elementary mathematics: Teachers' understanding of fundamental mathematics in China and the United States*. New York, NY: Routledge.

Mangels, J. A., Butterfield, B., Lamb, J., Good, C., & Dweck, C. S. (2006). Why do beliefs about intelligence influence learning success? A social cognitive neuroscience model. *Social Cognitive and Affective Neuroscience, 1*(2), 75–86.

Marinoff, L. (1994). A Resolution of Bertrand's Paradox. *Philosophy of Science, 61*(1), 1–24.

Mayer, R. E. (1992). *Thinking, problem solving, cognition* (2nd ed.). New York, NY: W. H. Freeman.

Mayer, R. E., & Wittrock, M. C. (2006). Problem solving. In P. A. Alexander & P. H. Winne (Eds.), *Handbook of educational psychology* (pp. 287–303). Mahwah, NJ, US: Lawrence Erlbaum Associates Publishers.

Moser, J., Schroder, H. S., Heeter, C., Moran, T. P., & Lee, Y. H. (2011). Mind your errors: Evidence for a neural mechanism linking growth mindset to adaptive post error adjustments. *Psychological Science, 22,* 1484–1489.

National Council of Teachers of Mathematics [NCTM] (1989). *Curriculum and Evaluation Standards*. Reston, VA: Author.

National Council of Teachers of Mathematics [NCTM] (2000). *Principles and standards for school mathematics*. Reston, VA: Author.

Organization for Economic Cooperation and Development [OECD]. (2013a). *Technical report of the survey of adult skills (PIAAC)*. Retrieved from https://www.oecd.org/skills/piaac/_Technical%20Report_17OCT13.pdf.

Organization for Economic Cooperation and Development [OECD]. (2013b). *The survey of adult skills: Reader's companion*. Paris: OECD Publishing. Retrieved from: http://dx.doi.org/10.1787/9789264204027-en.

Organization for Economic Cooperation and Development [OECD]. (2017). PISA 2015 results (Volume V): Collaborative Problem Solving, PISA. Paris: OECD Publishing. https://doi.org/10.1787/9789264285521-en.

Polya, G. (1954a). *Mathematic and plausible reasoning—Volume I: Induction and analogy in mathematics*. Princeton, NJ: Princeton University Press.

Polya, G. (1954b). *Mathematic and plausible reasoning—Volume II Patterns of plausible inference*. Princeton, NJ: Princeton University Press.

Polya, G. (1957). *How to solve it: A new aspect of mathematical method*. Princeton, NJ: Princeton University Press.

Santos, L. (2017). The role of critical thinking in science education. *Journal of Education and Practice, 8,* 159–173.

Simon, H. A., & Newell, A. (1971). Human problem solving: The state of the theory in 1970. *American Psychologist, 26*(2), 145–159. https://doi.org/10.1037/h0030806.

Skinner, B. F. (1966). An operant analysis of problem solving. In B. Kleinmuntz (Ed.), *Problem solving: Research, method, and theory* (pp. 225–257). New York, NY: Wiley.

Soh, K. (2019). *PISA and PIRLS: The effects of culture and school environment*. Singapore: World Scientific.

Tukey, J. W. (1977). *Exploratory data analysis*. Reading, MA: Addison- Wesley.

Wai, M. Y. (2011). Further applications of mathematics: Explore Ptolemy's theorem and its applications. *School Mathematics Newsletter, 17,* 43–58.

Wenke, D., Frensch, P. A., & Funke, J. (2005). Complex problem solving and intelligence: Empirical relation and causal direction. In R. J. Sternberg & J. E. Pretz (Eds.), *Cognition and intelligence: Identifying the mechanisms of the mind* (pp. 160–187). New York, NY: Cambridge University Press.

Wilson, J., & Clarke, D. (2004). Towards the modelling of mathematical metacognition. *Mathematics Education Research Journal, 16*(2), 25–48.

Wong, N. Y., Marton, F., Wong, K. M., & Lam, C. C. (2002). The lived space of mathematics learning. *Journal of Mathematical Behavior, 21,* 25–47.

Wong, P. W. (2018). *STEM for parents*. Hong Kong: Shing Tao Publisher.

Young, A. E., & Worrell, F. C. (2018). Comparing metacognition assessments of mathematics in academically talented students. *Gifted Child Quarterly, 62,* 259–275.

Yu, C. H., Anthony, S, & Behrens, J. T. (1995, April). *Identification of misconceptions in learning central limit theorem and evaluation of computer-based instruction as a remedial tool*. Paper presented at the Annual Meeting of the American Educational Research Association, San Francisco, CA. (ERIC Document Reproduction Service No. ED 395 989).

Yu, C. H., DiGangi, S., & Jannasch-Pennell, A. (2008). The role of abductive reasoning in cognitive-based assessment. *Elementary Education Online, 7(2),* 310–322. Retrieved from http://ilkogretim-online.org.tr/index.php/io/article/view/1804/1640.

Yu, C. H., Lee, H. S., Lara, E., & Gan, S. G. (2018). The ensemble and model comparison approaches for big data analytics in social sciences. *Practical Assessment, Research, and Evaluation*, 23(17). Retrieved from https://pareonline.net/getvn.asp?v=23&n=17.

Chapter 5
Self-efficacy and Productive Failure

While some psychologists argue that an inflated ego can be detrimental to learning, and in contrast, humility can lead to openness and hard work, others counter-argue that lacking self-efficacy can also affect academic performance. Evidence from 2015 PISA and 2016 PIAAC indicates that the relationship between self-image and academic performance is nonlinear in nature. Specifically, learners can cease to improve when they become content with their current skill level. The productive failure approach implemented in the Hong Kong educational system is proposed as a possible remedy to break down the ego of learners, thus making room for improvement. As one professor at Education University of Hong Kong interviewed by the team put it, "Humility is the key to learning".

5.1 Self-esteem, Self-efficacy, and Self-enhancement for the Chinese

Kwan, Hui, and McGee (2012) examined three aspects of self-processes among the Chinese: self-esteem, self-efficacy, and self-enhancement. Self-esteem has long been considered a beneficial factor for mental health and productivity. It is thought to be positively associated with subjective wellbeing and negatively associated with depression and anxiety. It appears that Chinese students tend to demonstrate lower self-esteem as compared with their Western counterparts and to show similar levels as those of other Asians. However, the dialectical thinking style of the Chinese might not provide an accurate representation of the Chinese mind when Western assessment tools are used. Dialectical thinking refers to the balance between two polarities. In dialectics, two seemingly contradictory attitudes can coexist. To be more specific, although a healthy self-image and self-deprecating behavior may appear to be the opposite of each other, there can be a disassociation between implicit and explicit measures of self-esteem. To put it simply, a Chinese person might still have a fine self-image while saying humble words in social settings. This is a crucial point

C. H. Yu and H. S. Lee, *Creating Change to Improve Science and Mathematics Education*, https://doi.org/10.1007/978-981-15-3156-9_5

because Chinese mental health status as measured by conventional scaling might be misleading. The authors will revisit this point in the result section of data analysis.

Self-efficacy is defined as people's beliefs about their ability to control their own pace and level of functioning, along with the events that affect their lives. Two main aspects have been examined: general self-efficacy and domain-specific efficacy. General self-efficacy has been positively associated with one's personal accomplishment, financial success, psychological wellbeing, and life satisfaction. Chinese persons are found to have lower levels of general self-efficacy as compared to their non-Chinese counterparts. However, general self-efficacy may be less significantly associated with wellbeing for Chinese persons than for Westerners. Domain-specific self-efficacy has been thought to be positively related to one's task performance. Results are inconclusive in terms of comparisons of levels of domain-specific self-efficacy between the Chinese and their counterparts from other cultures. Last, mixed results exist in research regarding whether the Chinese self-enhance or self-efface. In addition, there is variance in terms of levels of self-enhancement among Chinese sub-groups. Interestingly, the path leading from self-efficacy to symptoms of depression was significantly stronger for American youths than for Hong Kong youths. Specifically, a strong ego could make a person less resilient against adversity, thus increasing the chance of suffering from depression. (Chen, Chan, Bond, & Stewart, 2006).

Overall, previous findings of the literature review suggest that Chinese and Westerners differ in levels of self-esteem, self-efficacy, and self-enhancement bias. Additionally, studies have pointed out that sociocultural factors could play an important role in the malleability of the Chinese self-processes. Finally, Chinese self-concept has important implications for maintaining one's mental health, interpersonal relationships, and productivity. All these suggest that the study of the Chinese self-concept is a research direction worthy to be pursued.

5.2 Debate in Self-efficacy

A few decades ago, the social psychologist Bandura (1977, 1986, 1990, 1991, 1995, 1997) revolutionized the system of education by stressing the importance of self-efficacy in learning. According to the theory of self-efficacy, one's belief in his or her own ability can lead to a successful accomplishment of a task or fulfillment of a goal. Hence, his theory significantly contributed to the development of positive psychology, which focuses on and nurtures positive aspects of human life, such as happiness and wellbeing (Bandura, 2008). Among Bandura's many theories, self-efficacy has received much attention in the field of social and educational psychology (Williams & Rhodes, 2016). Self-efficacy is the belief one has in his or her abilities to produce certain levels of performance and/or completion of a task (Bandura, 1977). This belief can determine how people think, feel, motivate, act, and react.

However, Bandura (2012) also emphasized that self-efficacy is not an omnibus trait that is invariant across all facets of human experience. Some educators have misconstrued the notion of self-efficacy, the notion of boosting learners' self-images regardless of their performances. Often, students are given positive feedback irrespective of the actual quality shown in their academic outcomes. Protagonists of this strategy argue that uplifting a student's perception of self-image is not only a "good" intention but also projects the possibility where consistent encouragement leads to long-term betterment. In contrast, others insist that such an approach is a misshapen and an awfully misused example of Bandura's self-efficacy theory. Bandura (2012) himself had explicitly warned people of the danger of falsifying the true meaning of human self-efficacy: "If people experience only easy successes, they come to expect quick results and are easily discouraged by setbacks and failures. Resilient self-efficacy requires experience in overcoming obstacles through persistent effort. Resilience is also built by learning how to manage failure so that it is informative rather than demoralizing" (p. 13). In short, the growth of one's self-efficacy should be faithfully accompanied by realistic, vigilant evaluations and deliberate self-reflection on one's ability and performance. Lacking such a base, it is not surprising to see how the "Everyone gets a trophy" syndrome became questionable. Specifically, US educators are concerned with grade inflation, as evidenced by the fact that today 43 percent of all US college students receive "A" grades, while 86% of private school students receive "B" grades or higher (Sigman, 2012). Twenge (2010) argued that the "Everyone gets a trophy" mentality fails to build true self-esteem; instead, it leads to the building of ego inflation and a sense of entitlement. Not surprisingly, some researchers have found that contrary to the promised results, boosting self-efficacy has been debilitating to actual performance (Holiday, 2016; Vancouver & Kendall, 2006; Vancouver, More, & Yoder, 2008; Vancouver et al. 2002; Vancouver, Thompson, & Williams, 2001).

In the following section, the role of self-efficacy in learning will be examined by utilizing international assessment data. It is important to point out that some of these data are tied to self-efficacy while some are related to self-concept. The construct "academic self-concept" necessitates clarification because the terms "self-esteem", "self-concept", and "self-efficacy" are used by researchers from different disciplines. Researchers in the field of personality and social psychology typically concentrate on "self-esteem" in a global fashion, meaning the overall perception of the self as a person (Koller & Baumert, 2006). On the other hand, educational psychologists tend to narrow down the focus into self-concept and self-efficacy. At first glance, academic self-concept seems similar to self-efficacy, but there is a subtle difference between the two factors. Academic self-concept refers to learners' knowledge and perceptions about themselves in an overall academic domain (Wigfield & Karpathian, 1991) whereas academic self-efficacy refers to learners' beliefs that they can successfully perform a particular academic task (Schunk, 1991). An obvious example of the former is: "I have always been good at science" (Ferla, Valcke, & Cai, 2009). A typical example of the latter is: "How confident you are about understanding graphs presented in newspapers" (Anderson, Milford, & Ross, 2009). Further, it is crucial to point out that while academic self-efficacy is based on a goal-oriented self-evaluation of one's ability, academic self-concept is tied to comparing oneself with one's peers as the reference group (Bong & Skaalvik, 2003).

As mentioned before, Vancouver and his colleagues found an inverse relationship between self-efficacy and performance. Nonetheless, some researchers reached the opposite conclusion by utilizing linear regression or/and structural equation modeling (Chevaliera, Gibbons, Thorpee, Snelle, & Hoskinsf, 2009; Marsh & Martin, 2011; Morony, Kleitman, Lee, & Stankov, 2013; Williams & Williams, 2010). Both camps may be right. Educators in Hong Kong believe that it is imperative to find the middle ground between one's confidence and modesty. A professor at the City University of Hong Kong said,

> Confidence is good, but from my Chinese perspective, Westerners are overconfident no matter whether they know things or not. In contrast, the Chinese are too humble and are too afraid to make mistakes. First, you have to have an accurate understanding of strengths and weaknesses.

A male professor at Lingnan University said, "[Regarding self-efficacy and modesty], there should be a balance. I would fall in the middle. The more we learn, the more we realize our ignorance. Confidence gives us the power to ask questions".

5.3 What International Assessments Reveal

Our interpretation of the data runs the same course with the idea of finding a balance between one's self-confidence and one's humility. In one part of a curvilinear relationship, the association between self-efficacy and performance is positive. However, after passing the inflection point, the relationship is reversed. If educators focus on only one trend of the curve, the model might fail to provide useful guidelines for education reform. In the following section, the data sets from the 2015 Programme for International Student Assessment (PISA) and the 2016 Programme for International Assessment of Adult Competency (PIAAC) were utilized to investigate how one's self-efficacy in science and self-reported readiness to learn are correlated with test performance in a nonlinear fashion. Uncertainty is an inherent property of any estimation, and thus for every participant OECD reported ten plausible test scores, known as plausible values (PV) (OECD, 2013). Wu (2004) suggested randomly selecting PVs for analysis because each PV is equally good for estimating the population characteristics, as each set forms a sampling distribution of the estimated student ability. Following the advice of Wu, this study used a randomly selected PV for both PISA and PIAAC.

5.3.1 PISA 2015

The Programme for International Student Assessment (PISA), as its name implies, is a large-scale cross-cultural assessment of student performance administered by the Organization for Economic Cooperation and Development (OECD) every four years. The target population of this assessment consists of 15-year old students from 65 OECD member nations. The assessment covers literacy, math, and science, with a different focus in each round. In 2015 the focus was on science learning and therefore data for science self-efficacy (SSE), science self-belief (SSB), and ambition, in general, were collected. Science self-efficacy refers to a student's perception of his or her ability to successfully complete a science-related academic task or to reach an academic goal in science. Self-belief, also known as self-concept, refers to the self-evaluation of one's general ability in a domain (Marsh & Martin, 2011). Ambition is measured by the degree of concurrence with the statement "I see myself as an ambitious person". Table 5.1 is a summary of average math and science test scores, science self-efficacy scores, self-belief scores, and ambition scores of the USA and seven Asian countries/regions.

The preceding table indicates that students in the US showed the lowest math and science scores, in comparison to their peers in Asia. However, their level of self-efficacy in science, self-belief, and ambition was higher than those of students in Asian countries/regions. Acknowledging that the conventional regression modeling assumes a linear relationship, the authors turned to nonlinear modeling instead to better investigate the underlying data pattern. Lambda smoothing, artificial neural network, and transformation regression modeling were employed for triangulation.

Lambda smoothing is a manual method in which the curve is adjusted in an interactive way. By adjusting the interactive Lambda smoother, the authors detected a curvilinear relationship between the science test performance and the science self-efficacy (see Fig. 5.1). After all, Bandura was right. As one's self-efficacy improves, one's science test scores also increase. However, the argument of Twenge and Vancouver was right as well. After one's science self-efficacy passes a certain threshold, one's science test performance begins to reduce. It is important to point out that this nonlinear pattern phenomenon is not confined to students in America. This pattern was also detected among Asian samples, which may suggest the possibility that this nonlinear relationship does not vary by culture, at least between East Asian countries and the US. Succinctly, this analytic model revealed a more accurate snapshot of the underlying relationship between self-efficacy and actual performance, proving itself to be a better tool than the traditional method for nonlinear phenomena. The result provides an explanation for the failure of a self-efficacy-focused educational strategy that has significantly deviated from Bandura's original intention.

The authors also employed artificial neural networks (ANN) for triangulation, which reports both visual displays and statistical figures. Neural networks, as the name implies, try to mimic interconnected neurons in animal brains in order to make the algorithm capable of complex learning for extracting patterns and detecting trends (Kuan & White, 1994; McMenamin, 1997). As this approach artificially mimics human neurons in computers, it is also named artificial neural networks. ANN is a form of machine learning, meaning that to a large degree the artificial intelligence algorithm, not the analyst, makes the decision. It is built upon the premise that real-world data structures are complex and nonlinear, and thus they necessitate complex learning systems. Unlike regression modeling that assumes linearity, neural networks can model linearity and thus typically outperform regression (Somers & Casal, 2009).

Table 5.1 Average 2015 PISA scores (PV) of the USA and Asian countries/regions

	Math scores		Science scores		Self-efficacy		Self-belief		Ambition	
Country/Region	Mean	SD	Mean	SD	Mean	SD	Mean	SD	Mean	SD
China	541.74	100.84	528.34	98.56	0.06	1.17	0.19	0.87	2.98	0.73
Hong Kong	550.55	88.48	525.60	79.58	−0.07	1.22	0.21	0.95	2.80	0.80
Japan	533.30	88.18	539.03	93.28	−0.46	1.22	−0.51	1.02	2.64	0.82
Macau	543.98	79.03	528.59	81.84	−0.03	1.12	−0.50	0.81	2.63	0.81
South Korea	523.91	99.97	514.75	95.00	−0.02	1.23	0.34	0.98	2.84	0.75
Singapore	557.08	95.75	545.95	104.60	0.07	1.14	0.42	0.94	3.00	0.79
Taiwan	539.20	103.79	530.85	99.85	0.19	1.19	−0.01	0.89	2.92	0.76
USA	**474.35**	87.92	**502.60**	98.04	**0.29**	1.29	**0.65**	0.95	**3.25**	0.72

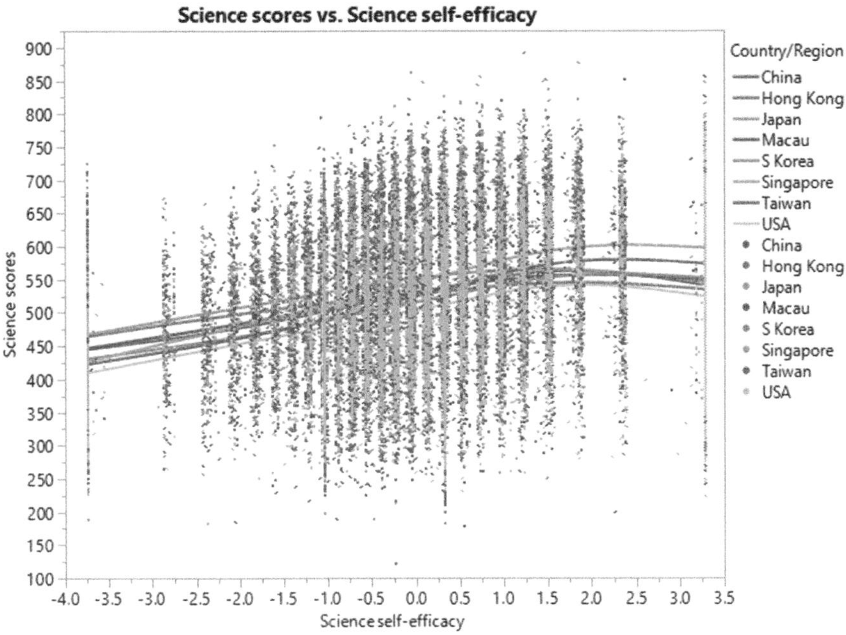

Fig. 5.1 Nonlinear model by Lambda smoothing

One may argue that Lambda manual smoothing of nonlinear curves is subjective, but ANN, which is capable of learning from the data, also suggests that while enhancing self-efficacy can improve performance initially, one's inflated self-efficacy could drag down one's actual performance. This pattern was common across cultures (see Fig. 5.2).

One major criticism against ANN is that it acts like a "black box", forcing humans to hand over judgment to the machine while we don't fully understand what is happening inside the process. To alleviate this problem, the authors programmed nonlinear models in SAS and made most of the decisions. Although there are over 20 transformation options, the authors used Penalized B-spine only because this approach tends to yield the optimal solution (Eilers & Marx, 1996). Figure 5.3 shows the result of the Penalized B-Spine fit. The main idea of PBS is to penalize complexity while controlling the smoothness of fit. By looking at Fig. 5.3, one can reach the same conclusion that Lambda smoothing and ANN suggested: positive self-efficacy can boost test performance to some extent, but an inflated ego can lead to the opposite outcome.

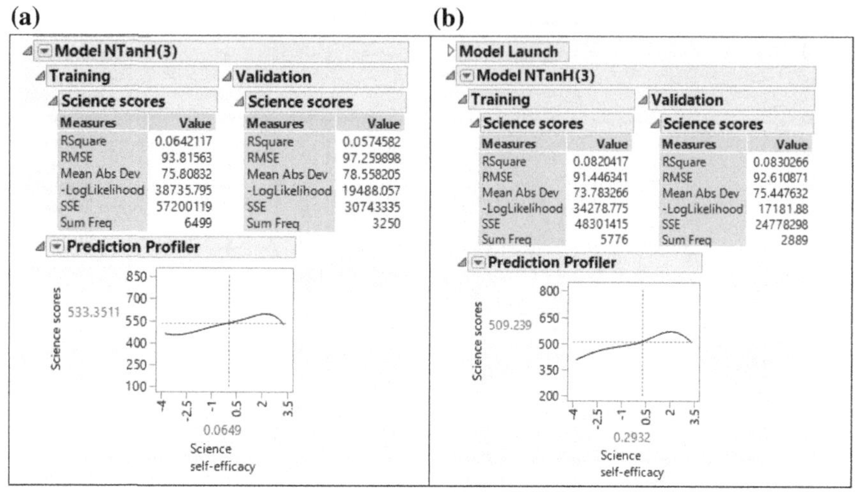

Fig. 5.2 **a** ANN of China's PISA data. **b** ANN of USA's PISA data

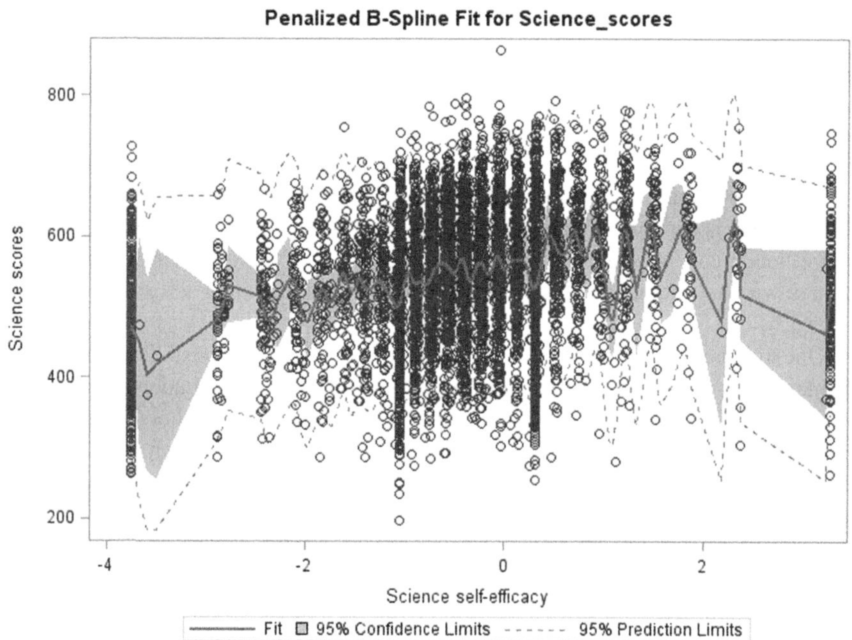

Fig. 5.3 Penalized B-spline model

Table 5.2 Average 2015 PIAAC scores of the USA and Asian countries ($n = 22,423$)

Country	Numeracy		Problem-solving		Readiness to learn	
	Mean	SD	Mean	SD	Mean	SD
Japan	289.47	43.82	295.31	43.77	2.96	0.81
Singapore	255.81	69.23	287.66	45.01	3.48	0.83
South Korea	262.28	46.14	283.19	37.24	2.86	0.94
United Sates	254.51	55.56	278.01	43.28	4.05	0.72

5.3.2 PIAAC 2016

The Programme for International Assessment of Adult Competencies (PIAAC) is an adult version of PISA. The age range of PIAAC participants is between 16 and 64. In PIAAC, three skill domains, including literacy, numeracy, and technology-based problem-solving, are assessed. Numeracy and digital problem-solving have become the focal point of US educators because like the PISA results, the US participants also fall behind their international peers in math- and science-related skills (OECD, 2012). This may imply that the US workforce is less competitive in today's technology-driven economy. In addition to test scores, PIAAC also collected data regarding self-reported readiness to learn. Readiness to Learn (RTL) is defined as the tendency to learn new things, to relate information to prior knowledge and life situations, as well as to engage in problem-solving and information-seeking behaviors. Table 5.2 is a summary of the average numeracy scores, digital problem-solving scores, and readiness to learn scores of the US and three Asian countries: Japan, Singapore, and South Korea (Hong Kong did not participate in 2016 PIAAC).

As shown in Table 5.2, American participants report the highest level of readiness to learn, but their actual test performance in numeracy and digital problem-solving is lower than their counterparts in Asian countries. Figure 5.4 is even more revealing. The RTL distributions of Japan and South Korea are fairly normal while that of Singapore is slightly skewed. However, not only is the US distribution skewed towards high RTL values but also the mode is 5 (the highest possible value), meaning that the majority of the US participants view themselves as being extremely ready to learn.

5.3.3 Lambda Smoothing

When the relationship between numeracy and RTL is investigated through Lambda smoother, one can see a curvilinear association: Nonlinear patterns are consistently detected across cultures. Numeracy scores improve as RTL goes up, but after passing the inflection point, the line turns flat or goes down slightly. However, unlike inflated self-efficacy that could adversely affect performance, increasing perceived RTL has less adverse effects on an individual's performance (see Fig. 5.5).

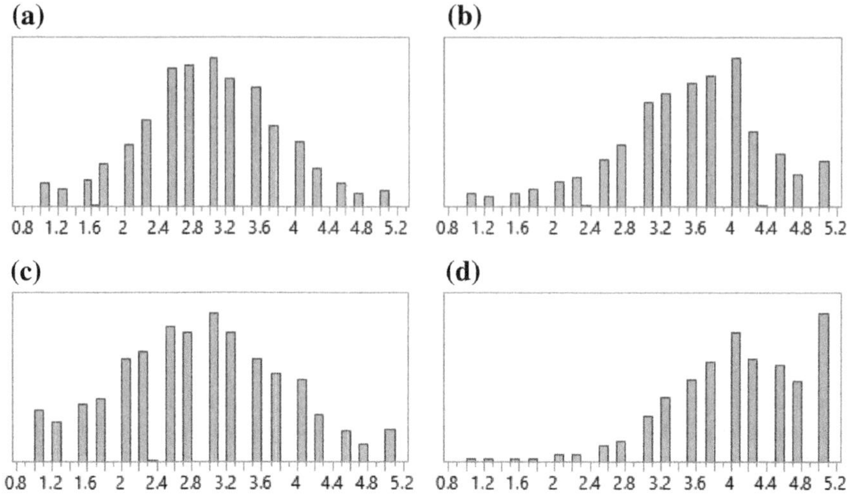

Fig. 5.4 **a** Distribution of RTL of Japan. **b** Distribution of RTL of Singapore. **c** Distribution of RTL of South Korea. **d** Distribution of RTL of USA

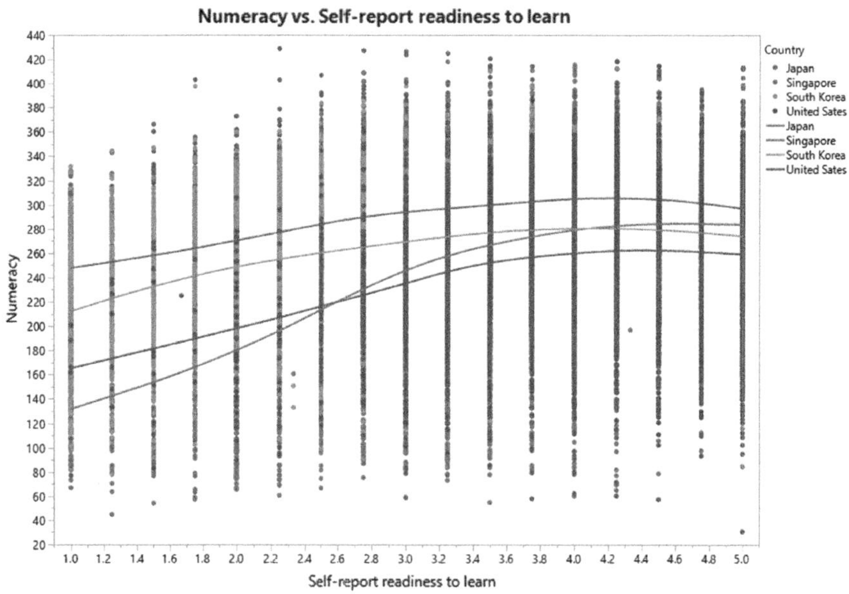

Fig. 5.5 Useful nonlinear model by Lambda smoothing

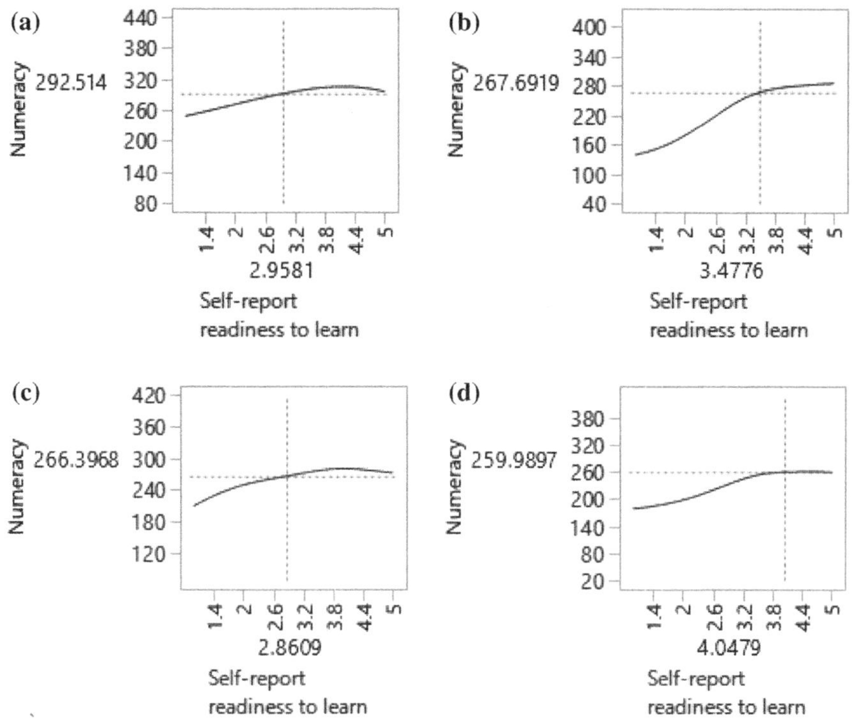

Fig. 5.6 **a** ANN of Japan. **b** ANN of Singapore. **c** ANN of South Korea. **d** ANN of USA

ANN was employed to verify the results. The findings of Lambda smoothing and ANN were congruent to each other. Once again, self-perception of readiness to learn and math skills go up hand in hand to a certain point. However, once it passes the threshold, one's inflated ego is no longer helpful in one's learning and could even drag down his or her performance (see Fig. 5.6).

Due to space constraints, only the Japanese sample was used for computing transformation regression models. Once again, the PBS model confirms the nonlinear relationship between self-assessment of readiness and actual performance (see Fig. 5.7).

5.4 What's Next: Productive Failure

5.4.1 Productive Failure in Education

Linear modeling is convenient, but it often fails to capture the essence of reality, resulting in misguided decisions. Whether boosting self-efficacy is the key to one's

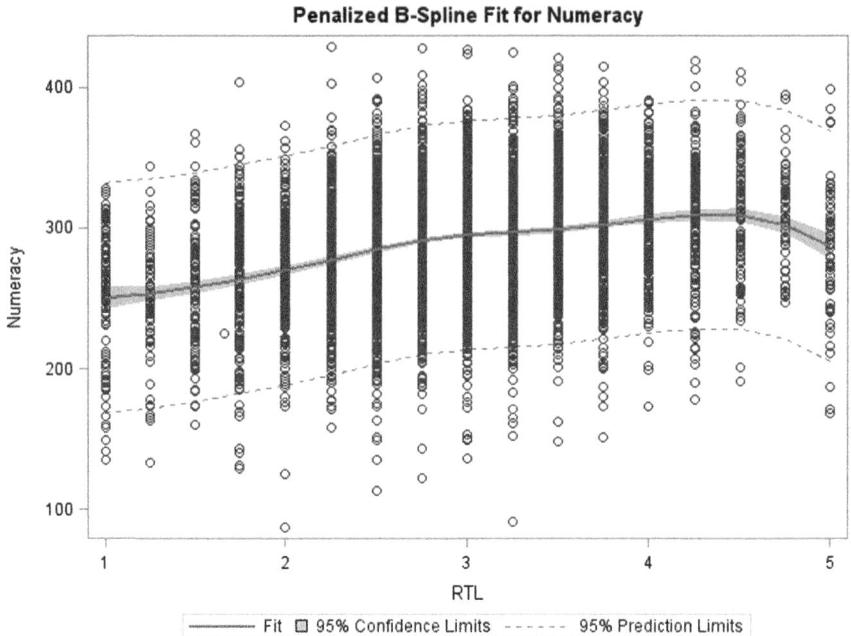

Fig. 5.7 Penalized B-Spline model of Japanese PIAAC data

better learning outcome has become a pressing issue in the field of education. However, the answer should not be simply dichotomous (i.e., Yes, it can/No, it cannot). It seems that previous research has overlooked the nonlinear character of this association, and has fueled unnecessary debates. For the past several decades, some US educators have attempted to boost academic performance by giving unconditional compliments and positive feedback to students, However, both PISA and PIAAC data show that there is a disconnect between self-perception of ability and actual performance. Asian cultures view modesty as a valuable trait, thus leading to a different pedagogical approach. As mentioned at the beginning, for Chinese people a fine self-image and a humble attitude can coexist, and thus evaluation using Western assessment tools might not provide a good indicator of Chinese mental health for Chinese people.

Interestingly, Hong Kong educators have been inspired by the way American innovators learned from their failures. For example, Lee and Shea (2018) wrote,

> Have you ever thought about where the Wright brothers, the aviation pioneers, found the solutions to the problem of flight? They acquired their first idea by observing how birds control their wings. As in many engineering processes, failures occurred. The Wright brothers built numerous prototypes in various sizes to test different components. They performed many trials and made repeated changes after each failure before they could claim even partial success. They had to develop the wings, the motor, the struts, the seating, the balance and the relationships between each of the components before the development of their first airplane (p. 118).

For a long time, Hong Kong educators have been systematically exposing students to manageable challenges and raising their self-awareness of misconception by employing the strategy of productive failure, which originated from Kapur, who was educated in America (2006, 2008, 2010, 2012, 2014, 2015, 2016). The idea is simple: in this learning mode, "students explored, discussed and solved problems related to the new concepts first in class even though they might come across failures, followed by consolidating the concepts and associated procedures" (Song & Kapur, 2017, p. 292). To put it another way, the instructor asks the learners to solve an ill-defined problem with ill-structured information while withholding guidance and feedback. It is expected that the learners will be frustrated, confused, and generate misconceptions, but at that point, the instructor corrects their misconceptions and consolidates their knowledge (Kapur, 2015; Song, 2018; Song & Kapur, 2017). To put it bluntly, the instructor sets them up to "fail". Empirical research shows that students learning science through productive failure gained a deeper understanding of conceptual knowledge and produced better solutions in problem-solving cases than their peers being assigned to the conventional method. Furthermore, those students who learned through productive failures showed more positive attitudes while facing challenges and also acquired a sense of ownership of their learning (Song, 2016, 2018). It is important to note that productive failure was not specifically designed for suppressing ego or promoting humility; rather, this approach was built upon the philosophy of discovery learning (Loibl & Leuders, 2018). Nonetheless, the authors argue that this method is applicable to the US setting.

5.4.2 *Resilience and Humility*

It is important to point out that one's excelling in science and math learning is tied to one's learning how to "fail well" or to be resilient to obstacles (International Association for the Evaluation of Educational Achievement, 2015). Specifically, according to the 2011 TIMSS data, 18% of Hong Kong students are considered disadvantaged yet academically resilient students. Both Singapore and Taiwan have 12%, compared with 8% in the US. OECD (2016) employed a different method for calculating the percentage of resilient students. According to the 2015 PISA data, 61.8% of Hong Kong participants were classified as "resilient students" and many other Asian countries that outperformed the US in PISA also had more resilient students than the US (see Fig. 5.8). Based on PISA data of resiliency, Cheung (2017) found that resilient youths were notably different from their non-resilient counterparts in terms of having greater self-regulatory skills and self-esteem.

When students have an inflated ego, a single incident of failure (e.g., one poor grade in a science class or one instance of negative feedback from the instructor) could be catastrophic enough to suppress their academic pursuit. Productive failure is a vaccine-like factor that can prepare students to eloquently normalize and facilitate the experience of failure in the study of science and math. Put bluntly, productive failure is an effective way of improving resilience.

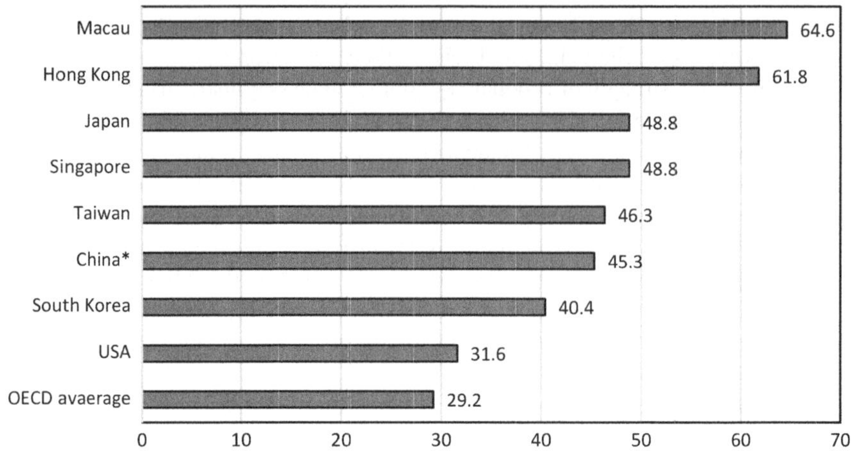

Fig. 5.8 Percentage of resilient students on PISA 2015 by country/region (*The Chinese sample includes Shanghai, Beijing, Jiangsu, and Guangdong only)

5.4.3 Learning from Failures in Organizations

This idea is by no means new in the US. For the last three decades, there has been a substantive increase in business research regarding how we can learn from failures rather than from successes (e.g., Atsan, 2016; Bazerman & Tenbrunsel, 2011; Cannon & Edmondson, 2005; Cope, 2011; Leoncini, 2015). After reviewing numerous cases, Madsen and Desai (2010) boldly asserted their findings that "organizations learn more effectively from failures than successes, that knowledge from failure depreciates more slowly than knowledge from success, and that prior stocks of experience and the magnitude of failure influence how effectively organizations can learn from various forms of experience" (p. 451). However, in business one would never deliberately create failure for learning purposes because the cost of such action would be too high and the consequences might be irreparable. There are countless examples of unintentional failures in business. For example, during the 1970s, in spite of a design flaw that could lead to a gas tank explosion in a rear-end collision, Ford Motor Company rushed the defective Pinto into production, resulting in the death of many drivers and passengers (Bazerman & Tenbrunsel, 2011). Needless to say, Ford Motor Company learned a hard lesson in engineering and business ethics and also had to work very hard to restore its image. Many years later Gerry Shaeen, a member of the board of directors for Ford Motor Company, said, "The ability to recognize and learn from one's own mistakes is the first step to becoming a great leader" (cited in Weinzimmer & McConoughey, 2013, p. 241). Unlike facing such a disastrous scandal, students can afford to make mistakes in the classroom because productive failure is managed by the teacher and the learners do not need to pay a high price. Interestingly, while Weinzimmer and McConoughey (2013) affirmed the value of learning from mistakes, they viewed ego as a major obstacle. They wrote,

"A successful leader needs to be confident, but when she starts to believe she is better than everyone else, her confidence has crossed the line, turning into self-absorption and arrogance—and bad things start to happen" (p. 242). Hence, humility and learning from mistakes must go hand in hand. This type of training should start earlier before personality, especially one's self-image, becomes stable across adulthood (Trzesniewski, Donnellan, & Robins, 2003).

5.4.4 Authentic Problems

Although teachers cannot directly transplant real business settings into the classroom, an ideal learning environment should be as authentic as possible so that students are better prepared for the future. One of the problems in math and science education, or even education in general, is that the exercises provided to the students are often too easy, unrealistic or non-normal in the real world. Take statistics as an example. In a typical exercise, the problem is clearly defined, all variables that are needed to solve the problem are included in the table, all data are cleaned, and the table is nicely structured in a row by column matrix. However, in the real world, the opposite is almost exclusively the case: the problem is ill-defined, some variables are missing, the data are messy, and there is no apparent structure. In conventional classrooms, learners might develop a false sense of competence by just following a step-by-step guideline to solve an easy problem. However, procedural based, hands-on exercises are vastly different from real science inquiry (Banchi & Bell, 2008). Using authentic learning tasks (e.g., uncertain problems in the real world) is a better pathway to transfer knowledge into authentic scenarios (Lai, Portolese, & Jacobson, 2017). In addition, when learners acquire concepts in a well-structured fashion, they might not understand why the concepts are organized in the way that they are (Schwartz & Bransford, 1998). Rote learning (i.e., following the textbook without knowing the rationale) is counterproductive to deeper learning. Prior research indicates that learners who were taught by the productive failure approach often developed better problem-solving abilities. This result has been verified in studies of 11th-grade physics problem-solving (Kapur, 2008), seventh-grade math problem-solving (Kapur, 2010; Song & Kapur, 2017), and ninth-grade math problem-solving (Kapur, 2012, 2014).

Interestingly enough, in a study of productive failure in STEM education (Truman, 2014), students reported a confidence level of 5.7 out of a 10-point Likert scale after being exposed to productive failure, whereas 11.5% of students were very confident about their results (8/10). After the experiment was completed, students became less confident in their ability ($M = 4.2$) and only 7.7% of students remained very confident (8/10). If subjective feelings (e.g., self-efficacy, self-concept) were used as the indicator of teaching effectiveness, then productive failure is doomed to fail. However, it was found that students made significant improvements in transferring problem-solving skill to different situations ($p = 0.0007$). This means that, although at first students subjectively felt less competent, they learned and applied more productively when they experienced failures.

5.4.5 Phases of Productive Failure

The strategy of productive failure is composed of two main phases. In Phase 1, students are encouraged to engage in the activity by raising questions related to the problem or the topic assigned, but no direct instruction is given. Next, students are asked to explore different resources and avenues in an attempt to solve the problem. Then, they are asked to explain the information gathered and the rationale behind their proposed solution. It is important to emphasize that this phase is not a means to an end. Rather, in productive failure, the process itself is as important as the outcome. In this exploratory process, prior knowledge possessed by the learner is activated, and this activation can help students to integrate new information at the problem-solving stage (Loibl & Leuders, 2018, 2019).

In Phase 2, the instructor explains why some proposed solutions are ineffective or inefficient, resulting in knowledge consolidation (Song & Kapur, 2017). At this point, the teacher must explain to the learners why certain concepts and solutions are erroneous. Prior research showed that when the instruction focused on correct answers only, the benefit (knowledge consolidation) of productive failure disappeared (Loibl & Leuders, 2019). As mentioned previously, productive failure is built upon the pedagogical philosophy of discovery learning. However, meta-analyses of over four decades of research on discovery learning indicated that discovery approaches are not necessarily superior to other teaching methods. One plausible explanation is that, without sufficient coaching discovery, learning could lead to confusion (Alfieri, Brooks, Aldrich, & Tenenbaum, 2011; Lazonder & Harmsen 2016). For instance, at Phase 2 when students are expected to fail, the teacher must stand beside them to correct their misconceptions, and thus turning failure to be a productive outcome. In this sense, productive failure is a form of guided discovery learning (Loibl & Leuders, 2018).

Productive failure can also be viewed as the highest form of inquiry in Banchi and Bell's taxonomy (2008). According to Banchi and Bell, there are four levels of inquiry-based science learning: 1. Confirmation, 2. Structured, 3. Guided, and 4. Open. In a typical confirmation inquiry, students are provided with the question and the guideline to find the answer. When the solution proposed by the student matches the modal answer, this confirms that the student has acquired the skill. At the next level, which is the structured inquiry, students are still provided with the question and tools. However, instead of mechanically following the procedure to find the solution, the learners are required to explain the answer. Thus, learners could understand the structural relationships between various concepts. Guided inquiry is more challenging because at this level only the question is given. The students need to find the proper procedure or method to obtain the answer. The last one, open inquiry, resembles productive failure. At this highest level, not even the question is clearly defined. Rather, the students are asked to act like real scientists, formulating the research question, designing and implementing a study, collecting data, and interpreting the results. Although the idea of open inquiry is promising, it is challenging for both teachers and students, and hence has rarely been implemented in science education (Häkkinen et al. 2017; Lakkala, Lallimo, & Hakkarainen, 2005). It is understandable that facing challenges, ambiguity, uncertainty, and even failure can be psychologically intimidating to most students. Therefore, to alleviate these issues, very often productive failure is integrated into collaborative project-based learning. In collaborative project-based learning, students work at real-world problems by experimenting with multiple solutions in small groups. Usually, it takes more than one person to come up with a viable solution, and therefore this situation necessitates collaboration. By doing so, learners have the opportunity to learn both hard (e.g., science and math) and soft (e.g., social) skills (Song, 2018).

5.5 Examples of Productive Failure in Hong Kong

Although the pedagogical strategy of productive failure originated in Singapore, this approach has been widely implemented by secondary school STEM teachers in Hong Kong. For example, in a ninth-grade lesson pertaining to enzymes, a science teacher at Ellen Yeung College (in Hong Kong, "college" is "secondary school") did not start the class session with a direct lecture. Rather, she broke the students into several groups and assigned them to collect information about enzymes using any available resources. Later, each group presented the information in a poster (see Fig. 5.9). At first glance, this approach may seem inefficient because students often face ambiguity and uncertainty during the process, which tends to increase their stress and require more time and effort. Nonetheless, students were able to develop problem-solving skills and an open mind as a result.

It is important to point out that productive failure should not be equally applied to all levels of learners. Rather, the vagueness of the instruction should vary from grade to grade. For instance, a Hong Kong teacher assigned less ambiguous science projects to lower level students while the upper division received "mystery assignments" in a contest (Grobler, 2018). The detail of these projects is illustrated as follows:

> S1 (equivalent to US Grade 7) students competed in a Battle of the Planes competition. Students were provided with twenty sheets of paper and some masking tape. They had to use all the paper to construct four different paper planes. They then had to stick five $1 coins on each plane. The plane that flew the farthest was declared the winner. Students were expected to do research before coming to the competition, but had to construct the planes on the spot. The requirement that they had to use all the paper and add the coins, which could make the planes off-balance, forced students to really think about the design of their planes (p. 24).
>
> S2 and S3 (equivalent to US Grade 8 and 9) students were required to build spaghetti bridges that could span 40 cm between two tables. Students used glue guns to connect the different pieces of spaghetti, but could not put glue along the length of any piece of spaghetti. The bridge that could carry the most weight without breaking was declared the winner. Again all the teams were expected to do research and plan what they were going to do before coming to the competition. This competition proved to be quite challenging. Even though students planned and designed their bridges, not all teams took the fragility of spaghetti into account" (p. 24).

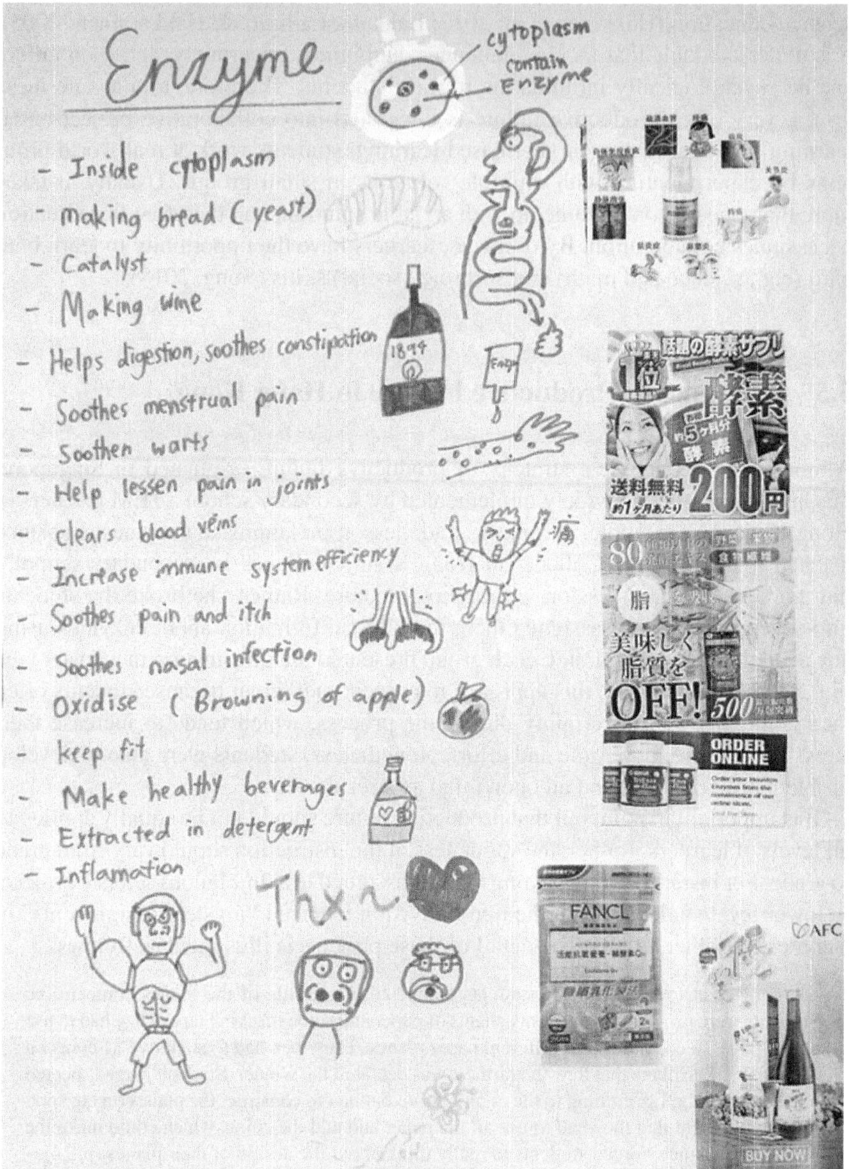

Fig. 5.9 Poster presentation on enzymes by Hong Kong students

> S4 and S5 (equivalent to US Grade 10 and 11) students competed in a mystery competition. They had no idea what they were going to do until the moment when they arrived at the competition. When they arrived, each team was provided with a container filled with random objects. The objects in each team's container were identical to those in the other teams' containers. The teams were then instructed to build boats that could travel the longest distance on the water without being pushed. Teams were also provided with iPads so that they could do research. Teams had to work together to plan and build their boats, do tests and adjustments before the final competition (p. 25).

Needless to say, these students, especially the S4 and S5 students who received the mystery assignment, encountered many unexpected obstacles and failures. The main point of this frustrating process was to learn from mistakes in a situation that mimicked real life. The nature of the mystery assignment is the antithesis of much current pedagogy that emphasizes clear structures and guidelines. Nonetheless, Grobler (2018) affirmed the value of mystery assignment, saying, "By trying different things and failing along the way, students will construct their own knowledge through their experiences" (p. 22).

5.6 Comparing Information Gap, and Productive Failure

Can we find any pedagogical strategy practiced by US educators that is similar to the productive failure approach advocated in Hong Kong? Perhaps the knowledge gap approach, also known as the information gap approach (Loewenstein, 1994), might be capable of achieving the same result as what productive failure intends to do. The root of the knowledge gap approach can be traced back to more than a century ago. US philosopher and psychologist James (1899) asserted that curiosity is the impulse towards better cognition, meaning that humans have a natural desire to learn about what we don't know. Based on this premise, Loewenstein (1994) and his followers utilized this type of intrinsic motivation to enhance educational attainment. For example, in a physics class the teacher showed a flat track and a U-shaped track, and then asked the students, "If I drop two balls on these two tracks simultaneously, which ball would reach the end first? There are three options: (a) The ball on the straight track will reach the end first. (b) The ball on the U-shaped track will reach the end first. (c) The race will end in a tie". The instructor attempted to build suspense by holding off giving the answer for a few minutes. After collecting responses from the students, the teacher explained that the ball on the U-shaped track could get to the end first because the downward motion of the ball increased its velocity (Freberg, 2019) (see Fig. 5.10).

Obviously, there is a common ground between the information gap approach and the productive failure strategy: in both cases, the instructors intentionally hide information from the learners. However, their similarity ends here. First, the information gap approach does not urge learners to try very hard, thus failing to require learners to reflect upon their misconceptions. The first author of this book observed some science classes in California that implemented the knowledge gap approach. Very

Fig. 5.10 YouTube video: High road low road race, potential kinetic energy Part 2–Homemade Science with Bruce Yeany

often some students gave no answer at all, while others made random guesses without processing the information. A moment later, the teacher gave the right answer and then the students copied the model answers into their worksheets. Needless to say, they did not recognize any knowledge gap, let alone reflect upon any misconception.

Second, it is noteworthy that the information gap approach is built upon an optimistic view of intrinsic motivation, which is prevalent in the Western intellectual tradition (Loewenstein, 1994). For instance, both Bentham (1789/1948, p. 34) and Burke (1757/1958, p. 31) viewed curiosity as a form of "appetite". In a similar vein, Freud (1915) also referred to curiosity as a "thirst for knowledge" (p. 153). By the same token, Piaget (1926/1960) regarded children as active and natural explorers. However, these positive views of motivation are contradicted by empirical data. According to a national survey, 69% of US K-12 teachers reported that students' lack of interest in learning was problematic (Yeager, Henderson, D'Mello, Paunesku, Walton, Spitzer, & Duckworth, 2014). Interestingly, the productive failure approach does not rely on a positive assumption of curiosity or interest. Many subjects in science and math are inherently boring and tedious. Rather than waiting for the students to become aware of their knowledge gap, the teacher must take the initiative to lead them into the acknowledgement of the gap. In summary, knowledge-seeking in the information gap approach is internally driven, whereas reflection upon misconception in the productive failure strategy is externally driven.

5.7 Discussion and Conclusion

In the last several decades, several good ideas had been proposed by educational researchers, such as discovery learning, open inquiry, and the information gap approach. Nevertheless, all of them have certain shortcomings. Unguided discovery learning tends to result in unresolved frustration and confusion; open inquiry is challenging to both teachers and students and therefore it is not widely implemented; information gap learning is built upon an optimistic view of curiosity and intrinsic motivation. Productive failure can remediate these limitations by providing students with guided discovery, collaborative project-based learning, and more importantly, turning their misconceptions and failures into positive learning outcomes. In the US, there is an idiom: "If you've never failed, you haven't tried hard enough to succeed". However, for several decades some educators have been trying to enhance learning by boosting learners' self-image. Very often, students learn math and science in a protective environment in the sense that explicit instruction is given in a well-assembled fashion. At the same time, teaching effectiveness is assessed by how learners feel about the class, the instructor, and themselves. As a result, learners might acquire a false sense of competency by earning a good grade easily. However, recent research indicates that perception is not equated with performance. Open-mindedness, which is crucial to learning, requires humility. It is the conviction of the authors that productive failure could be implemented in math and science education to improve both modesty and problem-solving skills.

References

Anderson, J., Milford, T., & Ross, S. (2009). Multilevel modeling with HLM: Taking a second look at PISA. In M. C. Shelley, L. D. Yore, & B. Hand (Eds.), *Quality research in literacy and science education: International perspectives and gold standards* (pp. 263–286). New York, NY: Springer.

Alfieri, L., Brooks, P. J., Aldrich, N. J., & Tenenbaum, H. R. (2011). Does discovery-based instruction enhance learning? *Journal of Educational Psychology, 103,* 1–18.

Atsan, N. (2016). Failure experiences of entrepreneurs: Causes and learning outcomes. *Procedia-Social and Behavioral Sciences, 235,* 435–442. https://doi.org/10.1016/j.sbspro.2016.11.054.

Banchi, H., & Bell, R. (2008). The many levels of inquiry. *Science and Children, 46*(2), 26–29.

Bandura, A. (1977). Self-efficacy: Toward a unifying theory of behavioral change. *Psychological Review, 84*(2), 191–215.

Bandura, A. (1986). *Social foundations of thought and action: A social cognitive theory.* Englewood Cliffs, NJ: Prentice Hall.

Bandura, A. (1990). Reflections on notability determinants of competence. In R. J. Sternberg & J. Kolligian Jr. (Eds.), *Competence considered* (pp. 315–362). New Haven, CT: Yale University Press.

Bandura, A. (1991). Self-regulation of motivation through anticipatory and self-regulatory mechanisms. In R. A. Dienstbier (Ed.), *Perspectives on motivation: Nebraska symposium on motivation* (Vol. 38, pp. 69–164). Lincoln, NE: University of Nebraska Press.

Bandura, A. (Ed.). (1995). *Self-efficacy in changing societies.* Cambridge, UK: Cambridge University Press. https://doi.org/10.1017/CBO9780511527692

Bandura, A. (1997). *Self-efficacy: The exercise of control.* New York, NY: Freeman.

Bandura, A. (2008). An agentic perspective on positive psychology. In S. J. Lopez (Ed.), *Positive psychology* (pp. 167–196). Westport, CT: Greenwood Publishing.

Bandura, A. (2012). On the functional properties of perceived self-efficacy revisited. *Journal of Management, 38,* 9–44. https://doi.org/10.1177/0149206311410606.

Bazerman, M. H., & Tenbrunsel, A. (2011). Ethical breakdowns: Good people often let bad things happen. Why? *Harvard Business Review, 89*(4), 58–65.

Bentham, J. (1789/1948). *Principles of morals and legislation*. New York, NY: MacMillan.

Bong, M., & Skaalvik, E. M. (2003). Academic self-concept and self-efficacy: How different are they really? *Educational Psychology Review, 15,* 1–40.

Burke, E. (1757/1958). *A philosophical enquiry into the origin of our ideas of the sublime and beautiful*. London, UK: Routledge and Kegan Paul.

Cannon, M. D., & Edmondson, A. C. (2005). Failing to learn and learning to fail (intelligently): How great organizations put failure to work to innovate and improve. *Long Range Planning, 38,* 299–319. https://doi.org/10.1016/j.lrp.2005.04.005.

Chen, S. X., Chan, W., Bond, M. H., & Stewart, M. (2006). The effects of self-efficacy and relationship harmony on depression across cultures. *Journal of Cross-Cultural Psychology, 37,* 643–658. https://doi.org/10.1177/0022022106292075.

Cheung, K. C. (2017). The effects of resilience in learning variables on mathematical literacy performance: A study of learning characteristics of the academic resilient and advantaged low achievers in Shanghai, Singapore, Hong Kong, Taiwan and Korea. *Educational Psychology, 37,* 965–982. https://doi.org/10.1080/01443410.2016.1194372.

Chevaliera, A., Gibbons, S., Thorpe, A., Snell, M., & Hoskins, S. (2009). Students' academic self-perception. *Economics of Education Review, 28,* 716–727.

Cope, J. (2011). Entrepreneurial learning from failure: An interpretative phenomenological analysis. *Journal of Business Venturing, 26,* 604–623. https://doi.org/10.1016/j.jbusvent.2010.06.002.

Eilers, P. H. C., & Marx, B. D. (1996). Flexible smoothing with B-splines and penalties. *Statistical Science, 11,* 89–121.

Ferla, J., Valcke, M., & Cai, Y. (2009). Academic self-efficacy and academic self-concept: Reconsidering structural relationships. *Learning and Individual Differences, 19,* 499–505.

Freberg, L. (2019, April). *Presidential address of Western Psychological Association: Harnessing curiosity for better learning*. Paper presented at 2019 Western Psychological Association Convention. Pasadena, CA.

Freud, S. (1915). Analysis of a phobia in a five-year-old boy. In *Collected papers* (Vol. 3, pp. 149–289). New York, NY: Basic Books.

Grobler, G. C. (2018). STEM in a day. *Hong Kong Science Teachers' Journal, 33,* 21–26.

Häkkinen, P., Järvelä, S., Mäkitalo-Siegl, K., Ahonen, A., Näykki, P., & Valtonen, T. (2017). Preparing teacher-students for twenty-first-century learning practices (PREP 21): A framework for enhancing collaborative problem-solving and strategic learning skills. *Teachers and Teaching, 23*(1), 25–41.

Holiday, R. (2016). *Ego is the enemy*. New York, NY: Penguin Group.

International Association for the Evaluation of Educational Achievement. (2015). Socioeconomically disadvantaged students who are academically successful: Examining academic resilience cross-nationally. Retrieved from https://www.iea.nl/fileadmin/user_upload/Policy_Briefs/IEA_policy_brief_Mar2015.pdf.

James, W. (1899). *Talks to teachers on psychology: And to students on some of life's ideals*. New York, NY: Henry Holt & Company.

Kapur, M. (2006). Productive failure. In S. Barab, K. Hay, & D. Hickey (Eds.), *Proceedings of the International Conference on the Learning Sciences* (pp. 307–313). Mahwah, NJ: Erlbaum.

Kapur, M. (2008). Productive failure. *Cognition and Instruction, 26,* 379–424. https://doi.org/10.1080/07370000802212669.

Kapur, M. (2010). Productive failure in mathematical problem solving. *Instruction Science, 38,* 523–550. https://doi.org/10.1007/s11251-009-9093-x.

Kapur, M. (2012). Productive failure in learning the concept of variance. *Instruction Science, 40,* 651–672. https://doi.org/10.1007/s11251-012-9209-6.

Kapur, M. (2014). Productive failure in learning math. *Cognitive Science, 38,* 1008–1022. https://doi.org/10.1111/cogs.12107.

Kapur, M. (2015). The Preparatory effects of problem solving versus problem posing on learning from instruction. *Learning and Instruction, 39,* 23–31.

Kapur, M. (2016). Examining productive failure, productive success, unproductive failure, and unproductive success in learning. *Educational Psychologist, 51*(2), 289–299. https://doi.org/10.1080/00461520.2016.1155457.

Koller, O., & Baumert, J. (2006). Self-esteem, academic self-concept, and achievement: How the learning environment moderates the dynamics of self-concept. *Journal of Personality and Social Psychology, 90,* 334–349.

Kuan, C., & White, H. (1994). Artificial neural networks: An econometric perspective. *Econometric Reviews, 13,* 1–91.

Kwan, V. S., Hui, C. M., & McGee, J. A. (2012). What do we know about the Chinese self? Illustrations with self-esteem, self-efficacy, and self-enhancement. In M. H. Bond (Ed.), *Oxford handbook of Chinese psychology* (pp. 279–294). Oxford, UK: Oxford University Press.

Lai, P. K., Portolese, A., & Jacobson, M. J. (2017). Does sequence matter? Productive failure and designing online authentic learning for process engineering. *British Journal of Educational Technology, 48,* 1217–1227.

Lakkala, M., Lallimo, J., & Hakkarainen, K. (2005). Teachers' pedagogical designs for technology-supported collective inquiry: A national case study. *Computers & Education, 45*(3), 337–356.

Lazonder, A. W., & Harmsen, R. (2016). Meta-analysis of inquiry-based learning: Effects of guidance. *Review of Educational Research, 86*(3), 681–718.

Lee, C. K., & Shea, M. (2018). Biomimicry-how nature inspires inventions. *Hong Kong Science Teachers' Journal, 33,* 118–122.

Leoncini, R. (2015). Learning-by-failing. An empirical exercise on CIS data. *Research Policy, 45,* 376–386. https://doi.org/10.1016/j.respol.2015.10.006.

Loewenstein, G. (1994). The psychology of curiosity: A review and reintrepretation. *Psychological Bulletin, 116*(1), 75–98.

Loibl, K., & Leuders, T. (2018). Errors during exploration and consolidation—the effectiveness of productive failure as sequentially guided discovery learning. *Journal of Mathematics Didakt, 39,* 69–96. https://doi.org/10.1007/s13138-018-0130-7.

Loibl, K., & Leuders, T. (2019). How to make failure productive: Fostering learning from errors through elaboration prompts. *Learning and Instruction, 62,* 1–10.

Madsen, P. M., & Desai, V. (2010). Failing to learn? The effects of failure and success on organizational learning in the global orbital launch vehicle industry. *Academy of Management Journal, 53,* 451–476.

Marsh, H. W., & Martin, A. J. (2011). Academic self-concept and academic achievement: Relations and causal ordering. *British Journal of Educational Psychology, 81,* 59–77. https://doi.org/10.1348/000709910X503501.

McMenamin, J. S. (1997). A primer on neural networks for forecasting. *Journal of Business Forecasting, 16,* 17–22.

Morony, S., Kleitman, S., Lee, Y. M., & Stankov, L. (2013). Predicting achievement: Confidence vs self-efficacy, anxiety, and self-concept in Confucian and European countries. *International Journal of Educational Research, 58,* 79–96.

Organization for Economic Cooperation and Development [OECD]. (2012). *Literacy, numeracy and problem solving in technology-rich environments framework for the OECD survey of Adult Skill.* https://doi.org/10.1787/9789264128859-en. Retrieved from http://www.oecd.org/skills/piaac/PIAAC%20Framework%202012–%20Revised%2028oct2013_ebook.pdf.

Organization for Economic Cooperation and Development [OECD]. (2013). *Technical report of the survey of adult skills (PIAAC).* Retrieved from https://www.oecd.org/skills/piaac/_Technical%20Report_17OCT13.pdf.

Organization for Economic Cooperation and Development [OECD]. (2016). *PISA 2015 results (Volume I): Excellence and equity in education.* Paris: Author. https://doi.org/10.1787/9789264266490-en.

Piaget, J. (1926/1960). *The child's conception of the world* (J. Tomlinson & A. Tomlinson, trans.). Totowa, NJ: Littlefield, Adams.

Schunk, D. H. (1991). Self-efficacy and academic motivation. *Educational Psychologist, 26,* 207–231.

Schwartz, D. L., & Bransford, J. D. (1998). A time for telling. *Cognition and Instruction, 16,* 475–522.

Sigman, M. (2012, June 18). When everyone gets a trophy, no one wins. *Huffpost.* Retrieved from http://www.huffingtonpost.com/michael-sigman/when-everyone-gets-a-trop_b_1431319.html.

Somers, M., & Casal, J. (2009). Using artificial neural networks to model nonlinearity: The case of the job satisfaction–job performance relationship. *Organizational Research Methods, 12,* 403–417. https://doi.org/10.1177/1094428107309326.

Song, Y. (2016). "We found the 'black spots' on campus on our own": Development of inquiry skills in primary science learning with BYOD (Bring Your Own Device). *Interactive Learning Environments, 24*(2), 291–305.

Song, Y. (2018). Improving primary students' collaborative problem solving competency in project-based science learning with productive failure instructional design in a seamless learning environment. *Educational Technology Research and Development, 66,* 979–1008. https://doi.org/10.1007/s11423-018-9600-3.

Song, Y., & Kapur, M. (2017). How to flip the classroom: Productive failure or traditional flipped classroom pedagogical design? *Educational Technology and Society, 20,* 292–305.

Truman, R. J. (2014). Productive failure in STEM education. *Journal of Educational Technology Systems, 42*, 199–214. https://doi.org/10.2190/ET.42.3.b.

Trzesniewski, K. H., Donnellan, M. B., & Robins, R. W. (2003). Stability of self-esteem across the life span. *Journal of Personality and Social Psychology, 84,* 205–220. https://doi.org/10.1037/0022-3514.84.1.205.

Twenge, J. M. (2010). *The Narcissism epidemic: Living in the age of entitlement.* New York, NY: Atria Books.

Vancouver, J. B., & Kendall, L. N. (2006). When self-efficacy negatively relates to motivation and performance in a learning context. *Journal of Applied Psychology, 91*, 1146–1153. https://doi.org/10.1037/0021-9010.91.5.1146.

Vancouver, J. B., More, K. M., & Yoder, R. J. (2008). Self-efficacy and resource allocation: Support for a nonmontomic discontinuous model. *Journal of Applied Psychology, 93,* 35–47.

Vancouver, J. B., Thompson, C. M., & Williams, A. A. (2001). The changing signs in the relationships between self-efficacy, personal goals, and performance. *Journal of Applied Psychology, 86,* 605–620.

Vancouver, J. B., Thompson, C. M., Tischner, E. C., & Putka, D. J. (2002). Two studies examining the negative effect of self-efficacy on performance. *Journal of Applied Psychology, 87*, 506–516. https://doi.org/10.1037/0021-9010.87.3.506.

Wigfield, A., & Karpathian, M. (1991). Who am I and what can I do? Children's self-concepts and motivation in achievement situations. *Educational Psychologist, 26,* 233–261.

Williams, D. M., & Rhodes, R. E. (2016). The confounded self-efficacy construct: Conceptual analysis and recommendations for future research. *Health Psychology Review, 10*(2), 113–128.

Williams, T., & Williams, K. (2010). Self-efficacy and performance in mathematics: Reciprocal determinism in 33 nations. *Journal of Educational Psychology, 102,* 453–466. https://doi.org/10.1037/a0017271.

Weinzimmer, L. G., & McConoughey, J. (2013). *The wisdom of failure: How to learn the tough leadership lessons without paying the price.* San Francisco, CA: Jossey-Bass.

Wu, M. (2004). Plausible values. *Rasch Measurement Transactions, 18,* 976–978.

Yeager, D. S., Henderson, M. D., D'Mello, S., Paunesku, D., Walton, G. M., Spitzer, B. J., et al. (2014). Boring but important: A self-transcendent purpose for learning fosters academic self-regulation. *Journal of Personality and Social Psychology, 107,* 559–580. https://doi.org/10.1037/a0037637.

Chapter 6
Blending Intrinsic and Extrinsic Motivation

Hong Kong's education system regards students' motivation, both intrinsic and extrinsic, as the engine for students to navigate throughout their entire learning process. A professor at a Hong Kong university noted that Hong Kong schools take careful notice of students' interests and motivation, as they are "important factors in rendering students' effective learning". The authors, however, believe that the construct of motivation must be examined in its corresponding cultural contexts and social settings. It has been established through multiple international comparisons that Chinese and other Asian groups from the Confucian heritage exhibit higher academic performance than their Western counterparts. However, the economic explanation often used to explain differences in academic performance among Western students clearly does not explain the higher performance for Chinese students, both at the national and family levels. Despite the fact that some Chinese have fewer economic resources, they still devote tremendous efforts to climb up the social ladder by studying. Hence, sociocultural and motivational factors have drawn interest among scholars (Hau & Ho, 2010).

First, it is common for Chinese students and their families to value high academic performance. Second, internationalized values and beliefs related to school accounts for success among these students more than merely personal interest. Extrinsic motivation and intrinsic motivation tend to occur simultaneously. Future research needs to move beyond the intrinsic versus extrinsic dichotomy in order to understand motivation within the Chinese cultural context. As discussed in Chap. 5, Chinese students do not tend to have overly high perceptions of their competence. Nevertheless, low competence perceptions may contribute positively to achievement, when their determination to succeed is high.

6.1 Debate on Intrinsic and Extrinsic Motivation

Motivation derived its meaning from the Latin word *movere*, which means "to move". A person with motivation acts toward an end that he or she projects as a satisfying

C. H. Yu and H. S. Lee, *Creating Change to Improve Science and Mathematics Education*, https://doi.org/10.1007/978-981-15-3156-9_6

goal (Ryan & Deci, 2000a). Traditionally, motivation is classified into two major categories: intrinsic motivation and extrinsic motivation (Eccles & Wigfield, 2002; Ryan & Deci, 2000a, b; Sansone & Harackiewicz, 2000). Intrinsic motivation is said to be the drive for action arising from internal factors, whereas extrinsic motivation is believed to be driven by external factors, such as utility value. Specifically applied to education, in the former case it would be more likely for students to be enthusiastic about learning mathematics and science because they perceive that knowledge of these subjects brings personal satisfaction; thus, they enjoy and appreciate the related work for its own sake (Gottfried, Fleming, & Gottfried, 2001; Taylor et al., 2014). This personal attainment may further develop into appreciating the beauty of nature and appeasing a deep curiosity through understanding the deeper structures of the universe (Taylor et al., 2014). On the other hand, students who are externally motivated may focus on the practical aspects and values of using math and science. A typical example is that engineering and programming are well-paid professions; thus the financial reward is worth the initial pain of taking these challenging majors. This classification is also applicable to many other domains. For example, an artist insisting upon "art for art's sake" is considered intrinsically motivated, while a commercial artist displaying his or her work in a for-profit gallery is regarded as extrinsically motivated. Prior research indicates that there is a seesaw relationship between intrinsic motivation and extrinsic motivation. If people care more about materialistic and monetary rewards, they typically care less about intrinsic values of social wellbeing (Music, 2014). By the same token, some educators refuse to give too many awards or incentives to students because they believe that students who are overly familiarized with being extrinsically motivated will be depleted of their intrinsic motivation (Sansone & Harackiewicz, 2000). Additionally, some researchers worried that extrinsic motivation can increase anxiety, thus affecting students' mental wellbeing. To be specific, externally motivated students who want to please their parents and teachers, and also to look good among their peers, may fear shame and disapproval from others if they fail to perform at the expected level. Eventually, they may become perfectionists who suffer from discouragement, a lack of confidence, and burnout (OECD, 2019).

Extrinsic motivation is also known as instrumental motivation. Unlike intrinsic motivation that is, in essence, self-driven, instrumental motivation is externally driven. To be specific, it is the desire to obtain something practical or rewarding, such as improving future career opportunities (Hudson, 2000). Prior research tends to downplay the role of instrumental motivation in learning. For example, a series of studies regarding second-language acquisition showed that instrumental motivation is a weak predictor of a student's determination to learn a second language (Gardner, 1983; Gardner, Lalonde, Moorcroft, & Evers, 1985; Gardner & Lambert, 1959, Gardner & Smythe, 1975). Based on their findings, Gardner and his colleagues concluded that instrumental motivation was appropriate for establishing short-term goals only, and was insufficient to sustain a task that requires perseverance. Nevertheless, Lukmani (1972) found a strong correlation between instrumental motivation and English proficiency ($r = 0.411$). It is important to note that these studies were conducted in the context of second-language acquisition, and therefore it is doubtful whether the findings could be generalized to learning science. It is the authors' conjecture that for East Asian students, instrumental motivation could strengthen their mentality of persistence toward learning math and science because the reward of acquiring sophisticated science skills is very high in the long term.

Intrinsic motivation is often tied to interest (Eccles & Wigfield, 2002; Eccles, Wigfield, & Schiefele, 1998; Pintrich & Schunk, 2002). Hence, triggering and maintaining interest is said to be the key to effective science and math education (Renninger, Nieswandt, & Hidi, 2015). However, prior research on the motivation mode of Asian students does not appear to be straightforward. Utilizing TIMSS data, Leung (2002) found that while Hong Kong, Singapore, South Korean, and Japanese students outperformed their Western peers in math, they held a negative attitude towards the subject. Another study shows that while students of Anglo background invest more effort into the subject matter of their interests, Chinese students distribute an equal amount of effort on all tasks, whether they are interested in the subject matter or not (d'Ailly, 2004). More importantly, among Chinese students, intrinsic and extrinsic motivation tend to coexist rather than to oppose each other (Salili, Chiu, & Lai, 2001).

Not surprisingly, the Hong Kong model is not built upon the conventional demarcation of intrinsic and extrinsic motivation. Kit-Tai Hau, a professor of Educational Psychology at the Chinese University of Hong Kong, and Irene Ho, a professor of Psychology at the University of Hong Kong (Hau & Ho, 2010), argued that although the original source of motivation is extrinsic, the gradual process of internalizing related values and goals can lead to a high degree of self-direction. An increasing number of studies pinpoint that the true relationship between intrinsic and extrinsic motivation is not necessarily mutually exclusive, but coinciding and interdependent, particularly from the longitudinal perspective. On the one hand, students may be extrinsically motivated by the expectations of their parents and the community, but on the other hand, they may also internalize the values of what they do (Cerasoli, Nicklin, & Ford, 2014; Covington, 2000; Hidi and Harackiewicz, 2000; Ryan and Deci, 2000b). Hence, in 2015, when OECD (2017) started to collect data regarding motivation, it used the construct "achievement motivation" without a clear demarcation between intrinsic and extrinsic motivation.

It is the conviction of the authors that the distinction between intrinsic motivation and extrinsic motivation is a false dichotomy. This stance hinges upon the assumption that no human action is fostered by a pure form of 100% intrinsic motivation or 100% extrinsic motivation. A science professor might declare that she is doing basic research without a commercial application because she is intrinsically motivated. However, is publishing the results in a peer-reviewed journal and eventually receiving a promotion a kind of external reward? A Microsoft programmer might be highly motivated by his six-figure income, but does he also enjoy a sense of self-actualization when he sees how the Windows OS revolutionizes the field of computing? Nevertheless, in this chapter, the authors still follow the convention by using the classical terms "intrinsic motivation" and "extrinsic motivation" in order to establish a common ground for dialogue and data analysis.

6.2 What TIMSS Data Reveal

Unlike South Korea, Japan, Taiwan, and Singapore, the economy of Hong Kong is service-centered, not technologically driven. Specifically, Hong Kong's economic composition by sector is as follows: services (34.7%), finance (25%), manufacturing (2.3%), and others (11.6%). Further, there is no research funding agency like the NSF or the NIH in Hong Kong. Consequently, finding a job directly related to math and science is difficult. In spite of the absence of a direct career path to fields related to math and science, Hong Kong students are still motivated to learn math and science. In the interviews, several Hong Kong students and parents told the first author that learning math and science is good because it could enhance logical reasoning, and this mode of reasoning can be well applied to all other domains (Yu & Wu, 2014). To put it bluntly, the academic success of Hong Kong students in these subjects is "in spite of" or "regardless of" their disadvantages, not "because of" their advantages.

One might jump to the conclusion that Hong Kong students are internally motivated because they appreciate the intrinsic value of math and science while there is no high-tech industry or funding agency for scientific research. However, careful scrutiny of TIMSS 2007, TIMSS 2011, and PISA 2015 data reveals a different story. Utilizing Jackknife regression for analyzing TIMISS data, it was found that in Hong Kong, the variable "science will help me in my daily life" is significantly and positively correlated with science test performance, but this association is absent from the US sample. Further, among US students, science test performance is adversely affected by enjoying science. Contrary to popular belief, enjoyment of science, which is considered an intrinsic motivator, is not a significant predictor of Hong Kong student performance in science. In addition, the variable "science is boring" was found to be a significant predictor of science test performance in both the US and Hong Kong (Yu, 2012) (see Table 6.1).

Table 6.2 shows the relationship between 2011 Grade 8 TIMISS math scores and the extent of agreement with the statement "I need math to learn other subjects". In the conventional sense, this variable is related to instrumental or extrinsic motivation because math is viewed as a means to an end. Interestingly, in Asia, including Hong

Table 6.1 Statistics of Jackknife regression models

Country/Region	Statement	Slope	SE[a]	*t*-ratio	*p* value
USA	I enjoy learning science	−8.20	1.74	−4.7	**<0.000005**
	Science is boring	3.02	1.14	2.6	**0.023268**
	Science will help me in my daily life	−0.48	1.70	−0.28	0.687924
Hong Kong	I enjoy learning science	2.52	3.88	0.65	0.524292
	Science is boring	−10.06	2.23	−4.52	**0.000022**
	Science will help me in my daily life	22.48	3.234	6.97	**<0.000005**

[a]Standard error

Table 6.2 Averages for math scores of G8 by agreement with "I need math to learn other subjects"

Country/Region	Agree a lot		Agree a little		Disagree a little		Disagree a lot	
	Average	SE[a]	Average	SE	Average	SE	Average	SE
Taiwan	645	(5.3)	620	(3.5)	609	(3.7)	544	(5.3)
Hong Kong	611	(4.9)	590	(3.8)	570	(4.6)	536	(7.8)
Japan	581	(5.2)	575	(2.8)	562	(3.4)	520	(6.7)
South Korea	650	(4.5)	618	(3.3)	592	(3.3)	563	(6.2)
Singapore	617	(4.2)	611	(3.9)	609	(4.4)	575	(9.5)
USA	**515**	(3.1)	**514**	(2.6)	500	(4.6)	476	(4.1)

[a]Standard error

Kong, whether students agree a lot or a little with the idea that math paves the way for learning other subjects can make a substantive difference in math performance. However, the math performance of American students remains virtually flat whether they agree with the statement a lot or a little.

A similar pattern is found in the relationship between math test scores and the extent of agreement with the statement "math will help in my daily life" (see Table 6.3). Asian students who agree a lot to this statement tend to have much better performance than those who agree only a little. But this difference is not observed among American students. Whether they agree with the statement a lot or a little, the average scores are exactly the same (512).

Regarding the statement "I need math to get the job I want", most Asian students appear to be motivated by the belief that studying math can lead to a desirable job (see Table 6.4). The only exception is that Singaporean students who reported a stronger level of agreement did worse than those who reported a lower level. Among US students, there is virtually no difference between those who agreed with the statement a lot or a little (514 vs. 512). Taking all of the above into consideration, TIMISS data reveals that while Asian students, in general, are motivated by external factors, US students are less likely to be extrinsically motivated.

Table 6.3 Averages for math scores of G8 by agreement with "math will help in my daily life"

Country/Region	Agree a lot		Agree a little		Disagree a little		Disagree a lot	
	Average	SE[a]	Average	SE	Average	SE	Average	SE
Taiwan	638	(4.2)	608	(3.3)	597	(4.9)	522	(6.9)
Hong Kong	604	(4.2)	581	(3.9)	561	(5.9)	540	(10.3)
Japan	581	(4.1)	571	(2.7)	564	(4.2)	520	(7.3)
South Korea	639	(5.0)	612	(3.0)	610	(3.8)	591	(5.5)
Singapore	615	(4.1)	607	(4.1)	613	(4.9)	585	(8.3)
USA	**512**	(2.9)	**512**	(2.8)	508	(3.5)	485	(4.7)

[a]Standard error

Table 6.4 Averages for math scores of G8 by agreement with "I need math to get the job I want"

Country/Region	Agree a lot		Agree a little		Disagree a little		Disagree a lot	
	Average	SE[a]	Average	SE	Average	SE	Average	SE
Taiwan	642	(4.8)	622	(4.3)	609	(3.4)	547	(4.9)
Hong Kong	604	(4.2)	586	(3.6)	577	(5.3)	515	(8.8)
Japan	585	(5.0)	572	(2.9)	564	(2.9)	528	(5.7)
South Korea	650	(3.6)	611	(3.2)	586	(3.5)	550	(5.9)
Singapore	609	(4.3)	612	(3.7)	627	(5.0)	575	(9.7)
USA	**514**	(2.8)	**512**	(3.2)	501	(5.1)	486	(4.7)

[a]Standard error

6.3 What PISA Data Reveal

PISA 2015 data concurs with the conclusions presented above. PISA also collects data pertaining to instrumental motivation, which is defined as the drive to learn mathematics and science based on the perception that math and science are useful to students and their future studies and careers (OECD, 2013). Figures 6.1 and 6.2 indicate that while there is a positive relationship between test performance in

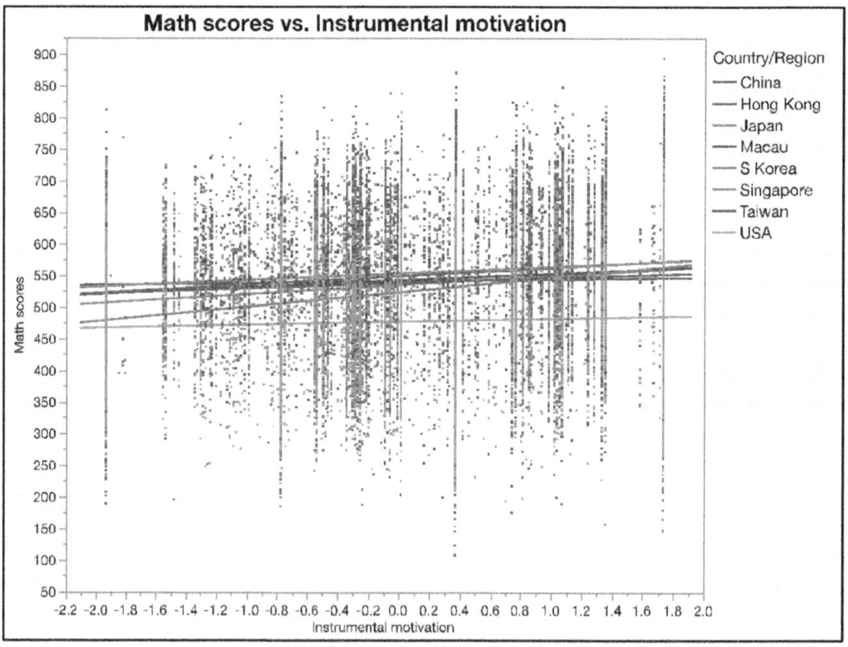

Fig. 6.1 Relationship between PISA math performance and instrumental motivation

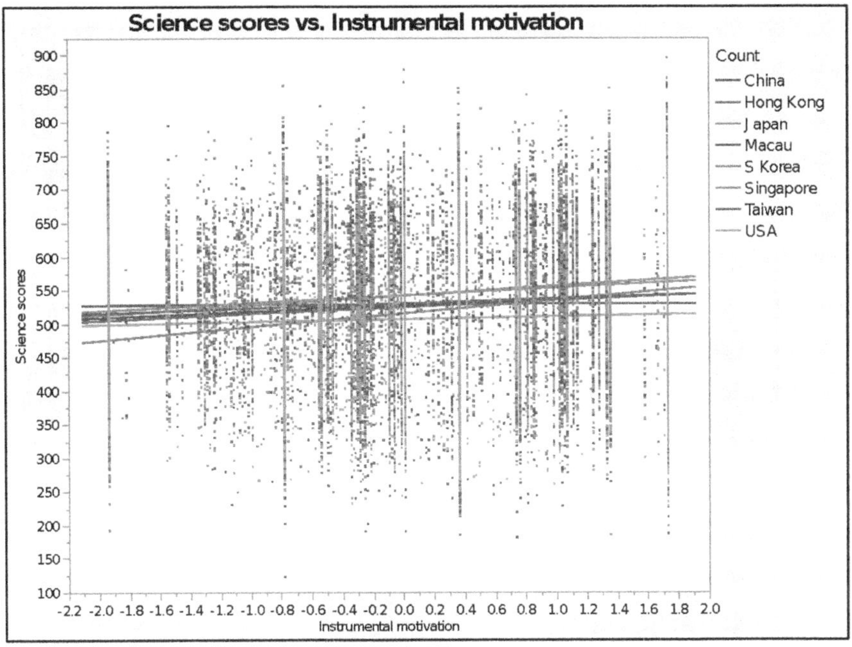

Fig. 6.2 Relationship between PISA science performance and instrumental motivation

math/science and instrumental motivation in Asia, the regression line of US students is nearly flat in both Figs. 6.1 and 6.2.

Conventional wisdom posits a seesaw relationship between intrinsic motivation and extrinsic motivation. While on average Americans are not extrinsically motivated, they can be intrinsically motivated. One of the popular counterarguments against modeling American education after Asia is that the Asian model, especially the Hong Kong one, forces every student to learn math and science, but America respects individual freedom; thus, US students who choose to learn math and science are very passionate. However, PISA data show that there is no seesaw association between intrinsic motivation and extrinsic motivation. The PISA survey has many variables related to intrinsic motivation and extrinsic motivation. For clarity of illustration, Fig. 6.3 shows only the scatterplot matrix of enjoyment of science (intrinsic), interest in broad science topics (intrinsic), and instrumental variables using observations of American and East Asian students ($n = 54{,}978$). It is apparent that the data cloud shows no pattern between instrument motivation and the two variables related to intrinsic motivation. Although the p values of all Pearson coefficients are significant, it is important to note that when the sample size is extremely large, any trivial correlation may be mistakenly reported as significant. Thus, pattern recognition by data visualization is a better method.

Because of overplotting, there is a possibility that the pattern is hidden inside the data cloud. To remediate this situation, the heat maps were utilized to detect a hidden

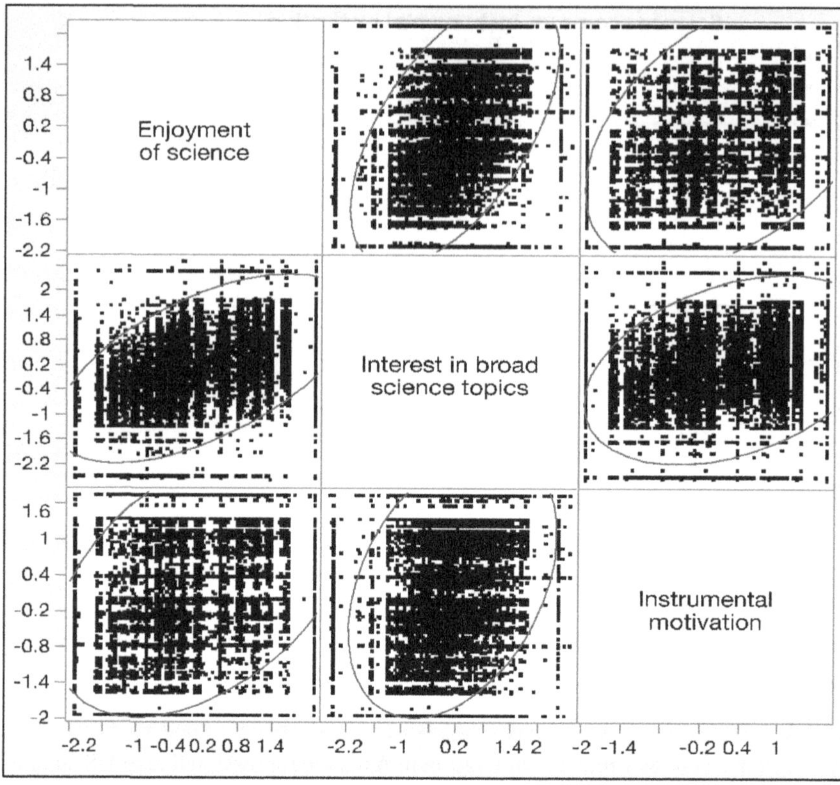

Fig. 6.3 Scatterplot matrix of variables related to intrinsic and extrinsic motivation

pattern. In a heat map, the frequency count of observations is depicted by colors. If there is a high concentration of observations, the hue is closer to the red end of the spectrum; otherwise, it is closer to the blue end of the spectrum. If there is a hidden relationship, one could see a red-pink ellipsoid in the scatterplot. Nonetheless, no such pattern was found. Figure 6.4 is one of the heat maps that was examined.

Nonetheless, the scatterplot in Fig. 6.5 suggests that there is a correlation between the two variables pertaining to intrinsic motivation (enjoyment of science and interest in broad science topics) because the data points form an upward ellipsoid.

Figure 6.6 is a scatterplot matrix of variables related to intrinsic and extrinsic motivation in the US sample. Again, no relationship is observed between intrinsic and extrinsic motivation, while there is a noticeable association between enjoyment of science and interest in broad science topics. Further, Figs. 6.7 and 6.8 indicate that there is a positive association between PISA science test performance and the two variables related to intrinsic motivation.

The preceding analysis confirms the hypothesis that intrinsic motivators are predictive of test performance. Now we can further investigate the difference between

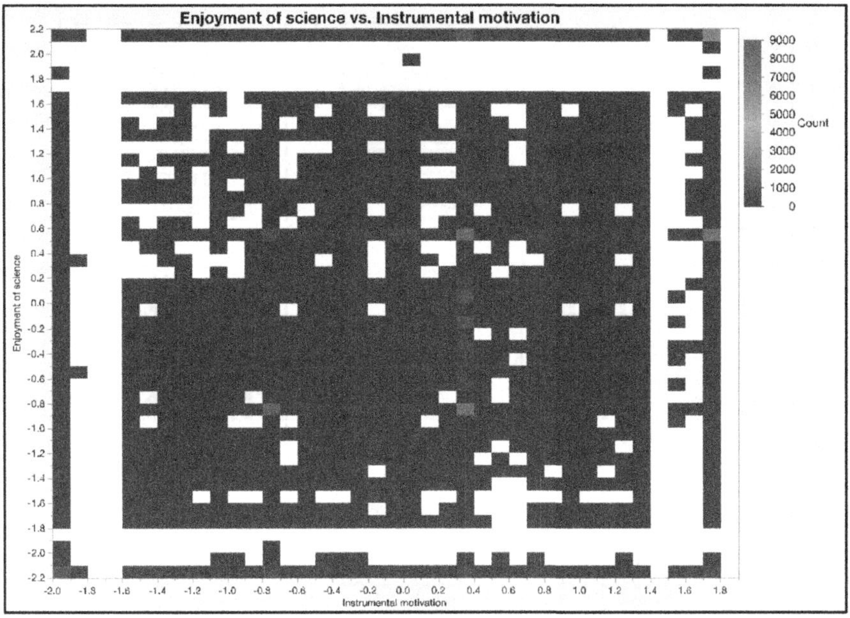

Fig. 6.4 Heat map of enjoyment of science and instrumental motivation

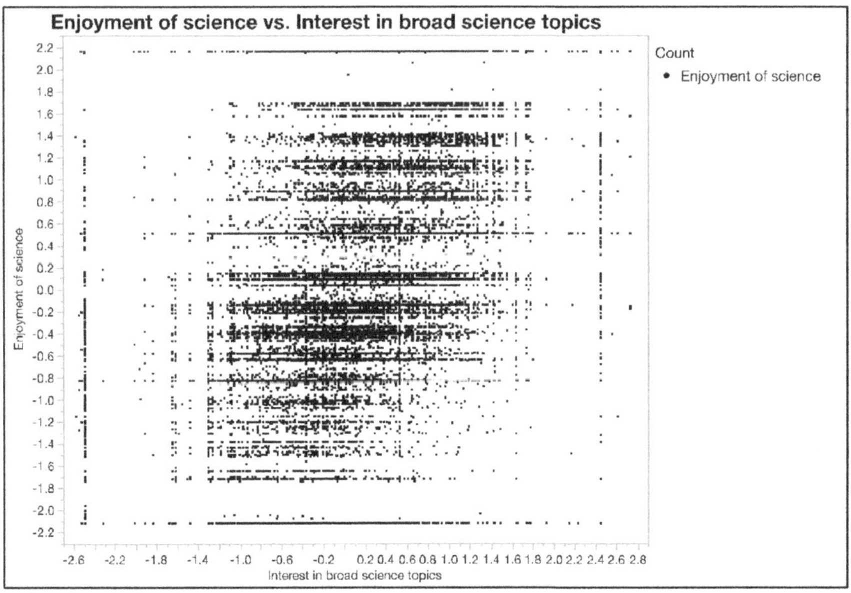

Fig. 6.5 Scatterplot of enjoyment of science and interest in broad science topics

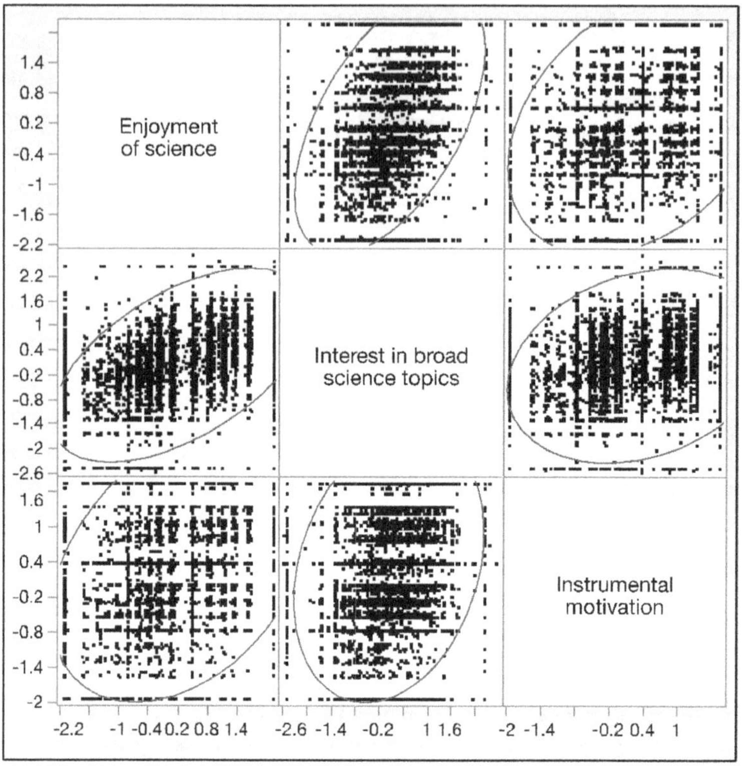

Fig. 6.6 Scatterplot matrix of variables related to intrinsic and extrinsic motivation in the US sample

Hong Kong and the US in terms of intrinsic motivation. Table 6.5 shows the descriptive statistics of test performance and intrinsic motivation, as measured by the enjoyment of science and interest in science. Interestingly, the mean scores of Hong Kong students' enjoyment of and interest in science are both higher than that of US students. However, as mentioned before, the demarcation between these two types of motivation is an artificial construct. When we turn to qualitative analysis, a different story emerges. This will be discussed in the next section.

In addition, there is a popular belief that because East Asian education, including the Hong Kong system, is exam-oriented and emphasizes external motivation, those students tend to suffer from academically related anxiety, thus decreasing their overall satisfaction with life. As a matter of fact, across all countries and regions that participated in PISA 2015 (except Macao), there is a significantly positive relationship between achievement motivation and satisfaction with life. Specifically, students in the top quarter of achievement motivation scored an average 37 points higher on the science test and reported 0.7 point higher life satisfaction (on a 10-point scale) than students in the bottom quarter (OECD, 2017).

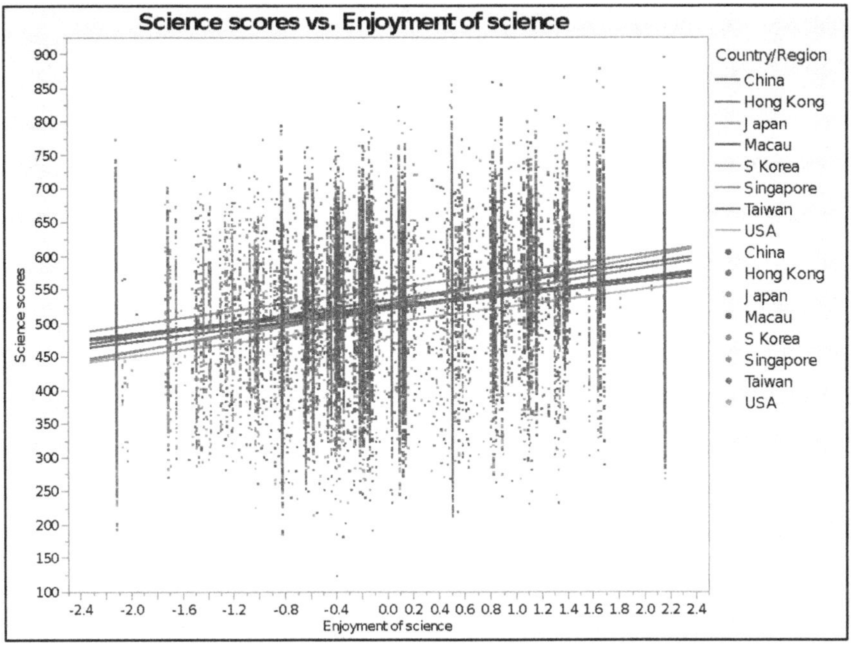

Fig. 6.7 Relationship between enjoyment of science and PISA science test performance

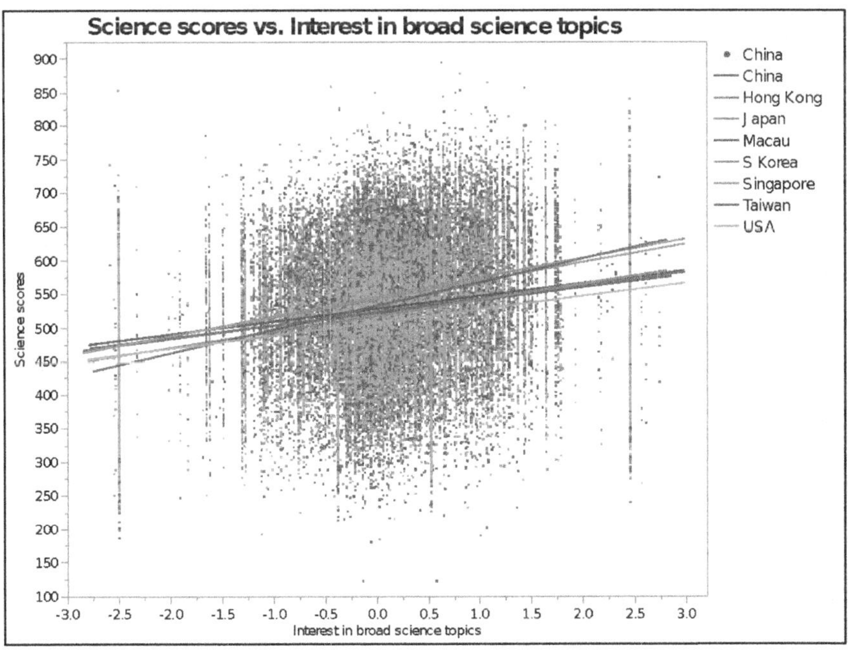

Fig. 6.8 Relationship between interest in broad science topics and PISA science test performance

Table 6.5 Test performance and intrinsic motivation scores of Hong Kong and US students

Country/Region	Math scores		Science scores		Enjoyment of science		Interest in science	
	Mean	SD	Mean	SD	Mean	SD	Mean	SD
Hong Kong	550.55	88.48	525.60	79.58	0.27	1.06	0.25	1.02
USA	474.35	87.92	502.60	98.04	0.24	1.07	0.04	0.97

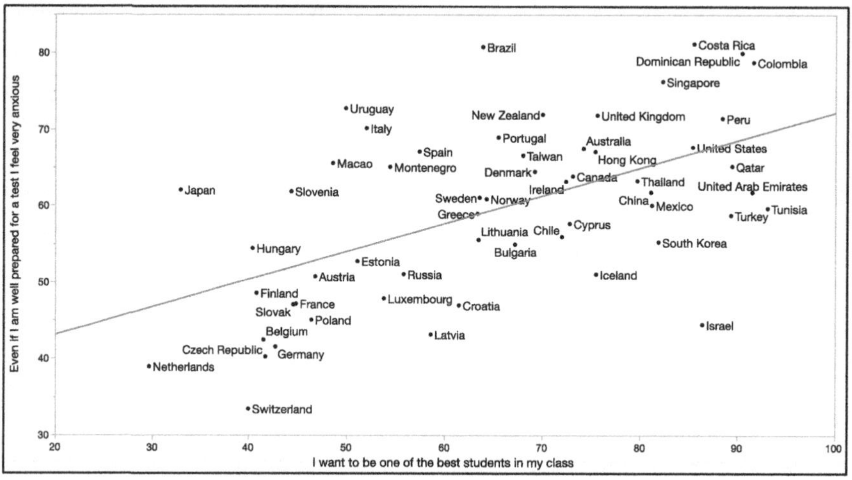

Fig. 6.9 Relationship between achievement motivation and test-related anxiety in 2015 PISA

While it is true that across all participating countries and regions there is a positive relationship between the percentage of students who agree with the statement "I want to be one of the best students in my class" and the percentage of students who agree with the statement "Even if I am well prepared for a test I feel very anxious" ($R^2 = 0.3$; $p < 0.0001$) (see Fig. 6.9), this relationship was absent in Germany, Hong Kong, and Iceland when individual-level data were examined (OECD, 2019).

More interestingly, while some Eastern countries or regions rank among the top in terms of the percentage of students who reported test-induced anxiety (Singapore = 5, Hong Kong = 14, and Macao = 17), others rank near the bottom (e.g., Japan = 24, China = 27, and South Korea = 37). By the same measure, some Western countries are among the top (e.g., U.K. = 8 and US = 12) whereas others are among the lowest (e.g., Iceland = 41 and Germany = 53). These data are available only for the 56 participating countries or regions; therefore, the cut-off between the top and the bottom is 18 (see Table 6.6). The GIS map in Fig. 6.10 shows a worldwide pattern. In this graph, countries or regions that have a higher percentage of students suffering from test-related anxiety are depicted in red, countries that have a lower percentage are depicted in blue, and countries in the middle are depicted in light purple. Interestingly,

Table 6.6 Ranks of test-related anxiety and achievement motivation in 2015 PISA

Country or Region	Rank of anxiety	Even if I am well prepared for a test I feel very anxious	Rank of motivation	I want to be one of the best students in my class
Singapore	5	76.3	11	82.3
UK	8	71.9	16	75.6
US	**12**	**67.7**	**10**	**85.4**
Hong Kong	**14**	**67.1**	**18**	**75.4**
Taiwan	16	66.6	26	68
Macao	17	65.6	43	48.6
Japan	24	62.1	55	32.9
China	27	61.8	14	81.1
OECD average		55.5		59.2
South Korea	37	55.3	12	81.9
Iceland	41	51.1	17	75.5
Germany	53	41.6	49	42.7

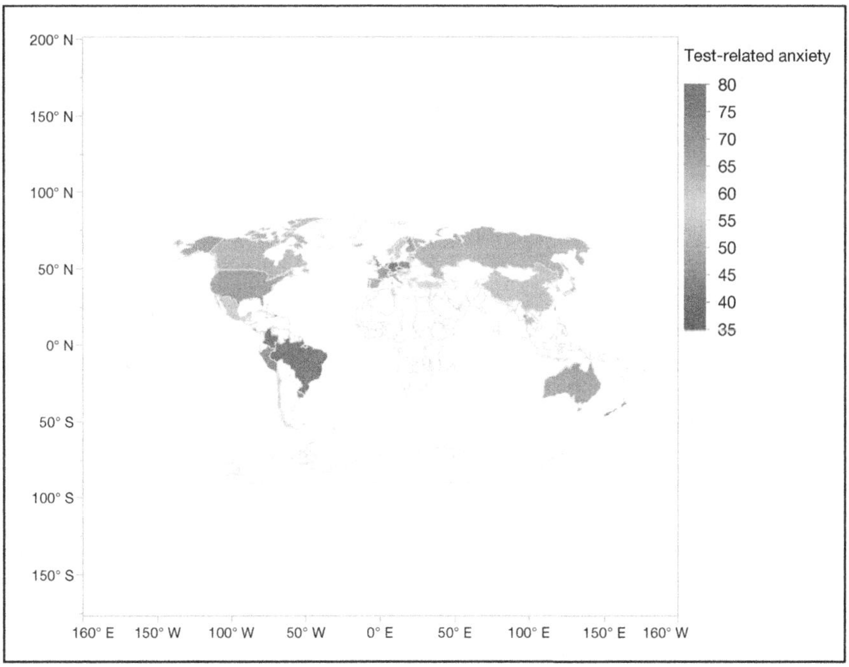

Fig. 6.10 GIS map of test-related anxiety

test-related anxiety seems to be more prevalent in North America, South America, and Australia, but less in Western Europe, Russia, and certain Asian countries.

It is noteworthy that while the US and Hong Kong have similar percentages of students who suffer from test-related anxiety (67.1% vs. 61.1%), the US has a higher percentage than Hong Kong of students who want to be one of the best (85.4% vs. 75.4%; see Table 6.6). Hence, using any stereotype in cross-cultural comparison is not advisable. The motivation approach in Hong Kong does not necessarily increase test anxiety among students.

6.4 Qualitative Analysis

Although the focus of this book is on high school students, it is important to point out that many university students develop their interest in math and science before commencing their postsecondary education. The book *Who says engineering is a fluffy discipline?* compiled by Kang and Kang (2015), alumni of Hong Kong University of Science and Technology, clearly indicated that many university students choose engineering as their major due to a mixture of intrinsic and extrinsic motivation. In this book, the authors collected stories of 46 engineering students or alumni of Hong Kong University of Science and Technology with the intention to debunking the myth among Hong Kong high school students, parents, and the general public that engineering is a meaningless discipline, with low admissions criteria and limited career options upon graduation. The idea that engineering is a less useful discipline might seem strange to American readers who live and work in a technology-driven economy, in which an engineering degree is a ticket to a high-paying career in Google, Cisco, Intel, and Apple. As mentioned before, Hong Kong's economy is service-oriented, not technology-based. Hence, it is not surprising that the public tends to conceive of selecting engineering disciplines as buying insurance.

In the book, the participants came from six engineering fields: chemical engineering, civil engineering, computer engineering, mechanical engineering, industrial and logistics engineering, and electrical engineering. The stories of the participants show that engineering is a highly applied discipline with a great deal of creativity involved. They think of engineering as an all-encompassing subject. For example, those participants in chemical engineering believed that this discipline is closely related to people's everyday life. From medicines to batteries to solar panels, all production requires knowledge in chemical engineering. Many professors of chemical engineering proudly tell their students: "Chemical engineers can do anything!" Although the chemical engineering industry is limited in Hong Kong, there is a greater world out there, whose door can be opened by knowledge in chemical engineering. Civil engineers are dedicated to the study of structures and designs. They plan, construct, and repair. Civil engineers build bridges that cross oceans as well as skyscrapers. They construct cities and environments that are healthy for human living. Further, computer engineers do not have to be lonely programmers. They can seek to understand the real needs of people and design products that suit such needs. They can use their expertise to accomplish things in the business world. And mechanical engineers protect the public by making sure that systems operate in a safe manner. Participants in industrial and logistics engineering believe that engineering can meaningfully assist the financial industry. They also believe that it is more important to find what one wants to do than simply to choose a major. By the same token, electrical engineers find great passion in studying the subject and applying it to their respective careers upon graduation. Their stories indicate that intrinsic and extrinsic motivation blends together seamlessly. At first glance, learning a skill is a practical reason, which seems to fall into the category of extrinsic motivation; however, one needs to have a passion for serving the public in order to choose a major that promises less in the way of financial rewards.

One of the assumptions of force-option survey items is that the respondents can truthfully report what they have in mind. In reality, our conception is fuzzy. Consider this example: Fara Lu was a Hong Kong university student majoring in chemical engineering. She said that when she selected chemical engineering as her major,

> I don't have a clear idea of what chemical engineering is about, or how this discipline would lead me to a profession in the future... During the learning process in the university, I gradually like this subject matter. One of the main points of chemical engineering that I like is that it simplifies many complicated problems. At the beginning, many core subjects of chemical engineering, such as the law of thermodynamics and transmission, seemed very difficult. But after I got it, I found that these complex phenomena could be epitomized into one or two very clean formulas. Conversely, I can start from a simple formula to deduce many other formulas that can be applied into various situation...Another reason that I like chemical engineering is: this discipline has a wide variety of applications. Many professors of chemical engineering proudly told us, 'Chemical engineers can do anything.'... I remember that after the semester was over I went to shopping for cosmetics. I was familiar with the ingredients of those products. I could even tell what their functions are. I had a sense of achievement (Kang & Kang, 2015, pp. 40–42).

How should her mode of motivation be classified? Probably it is a mixture of intrinsic and extrinsic motivation. On the one hand, she appreciates the beauty of mathematical structure and the logical structure of deduction. On the other hand, she also likes the subject matter because it can "do anything", including applying it to her everyday life (shopping for cosmetics). More importantly, in the beginning, she did not know what she wanted to do.

Consider another example: Louis Lam was another chemical engineering major. In his freshman year, his professor asked him this question: "Do you know the difference between a scientist and an engineer?" For those who understand the psychological theory of motivation and history of science, the answer is obvious: a scientist is a researcher who is intrinsically motivated in the research endeavor. Usually, she is involved in basic or theoretical research that has no immediate or practical applications. In other words, a theoretical researcher is concerned with satisfying her curiosity by seeking the scientific truth in forms of theories that comprise explanatory laws. In contrast, an engineer or an evaluator conducts inquiries assigned by someone else for finding the worth of particular products (Glass, 1972). However, Lam said that he could not answer the question at that moment. He didn't understand what constituted engineering. Nonetheless, as he grew older, he applied his engineering knowledge to many substantial projects. He said, "Engineering emphasizes logical reasoning and seeks for feasibility rather than a possibility. Emphasizing parallel and sustainable development gives people hope and courage for change. These abilities and resources can be well applied to both work and all other aspects of life… Career does not represent the whole life" (Kang & Kang, 2015, pp. 68–70). Again, it is difficult to fit his mode of motivation into either intrinsic or extrinsic motivation. Notably, Lam also did not know what he would do about chemical engineering at the beginning.

6.5 Discussion and Conclusion

In summary, large-scale international assessment data, including TIMSS and PISA, suggest that Hong Kong and other Asian students are motivated by so-called extrinsic factors while US students, to some extent, are immune to this type of motivation. The lack of significant association between the perceived usefulness of science in daily life and test performance among US students might be alarming, but it is by no means a new phenomenon. According to Feinstein (2011), there is little evidence that the prevalent practices of the US science education have an impact on the scientific applications in American students' daily life. Furthermore, there is also a lack of evidence that science taught in the US school helps students lead happier, more successful, or more politically savvy lives. In response to the broken link between science education and the daily applications of science, Feinstein (2011) suggested that science teachers should adopt a scenario-based approach by firmly embedding practical questions within the relevant context of the learners, instead of starting with abstract concepts. Specifically, Feinstein asserted, "Science education can help people solve personally meaningful problems in their lives, directly affect their material and social circumstances, shape their behavior, and inform their most significant practical and political decisions" (p. 169).

As indicated in TIMSS data, among the US students science test performance is adversely affected by enjoying science, while this variable is not found to be significant among Hong Kong students. In addition, the variable "science is boring" was found to be a significant predictor of American science test performance. At first glance these findings are contradictory and thus await further investigation. One plausible explanation is that whether a subject matter is boring or not contributes just a portion to the construct "science enjoyment". Many subjects include boring components (e.g., tedious calculations), and one can still enjoy science in spite of boredom, as indicated in our qualitative analysis. When the focus is directed to the construct "science enjoyment", to a certain extent our findings concur with those of others. For a long time, most innovative US educational reforms have centered on the development and application of multimedia technology, resulting in a more enjoyable learning experience. However, using all PISA 2006 data across 57 countries, Perez and Cromley (2009) found that the association between science scores and science enjoyment was the strongest among high achievers only.

Additionally, PISA data show that there is no seesaw relationship between intrinsic and extrinsic motivation. Specifically, Hong Kong students who are extrinsically motivated also show a higher mean score in "interest in broad science topic" and "enjoyment of science" than their American peers. Furthermore, it is important to point out that in a survey the participants are forced to choose a mutually exclusive position, but qualitative reports of certain Hong Kong students tell a different story. First, the student may not even able to tell how he or she was initially motivated. Second, as the student becomes more familiar with the subject matter, he or she becomes both intrinsically and extrinsically motivated. Finally, it is not necessarily true that the so-called achievement-oriented countries and regions in East Asia are more likely to generate a culture of high academic pressure than their Western counterparts. Indeed, there are more students in the US who reported suffering from test-related anxiety than their Hong Kong peers.

What is the implication for education reform in the US? It is very common for US teachers, professors, and parents to give this type of advice to children and youths: "Follow your heart. Don't do something that you don't have a passion for". In this sense, intrinsic motivation seems to be the primary motivator. As mentioned before, many US researchers believe that triggering and sustaining interest is crucial for successful math and science education. In order to make the content interesting, education becomes "edutainment", and usually technology is employed to make learning fun and exciting (Svencer, 2012). Paradoxically, the approach of triggering interest and making things fun turns out to be an extrinsic motivator. Specifically, students will not be motivated to learn math and science, which are inherently difficult and dry, and inevitably cause confusion and frustration. By contrast, Hong Kong educators do not tell students to follow their passion or to do something that they enjoy. The implicit assumption is that usually young children and youths, as indicated by the Fara Lu and Louis Lam, do not know what they would like to do in the future. Nonetheless, as Hau and Ho (2010) pointed out, students may start from extrinsic motivation regardless of enjoyment and interest, but as they mature, they can internalize the values of the discipline and become both intrinsically and extrinsically motivated. One professor who originally hails from the United States and now teaches at a university in Hong Kong shared a personal impression about his students:

> One major reason that I was interested in moving from the United States to Hong Kong was that I wanted to learn more about the Hong Kong educational system, which resulted in high performing students. My students are bright, reasonably motivated, and surprisingly interested in science (considering we have no science majors at the university).

Here, the professor did not specify the type of motivation underneath his students. Yet, he appreciates his students' desire and interest in learning science-related subjects, regardless of their majors. His impression may represent the majority of the students in Hong Kong. They are acutely clear about why and what they have to learn. They are driven by powerful motivations, whether extrinsic (such as exams, rank-focused competitiveness, and a high salary in their future career) or intrinsic (such as seeking personal accomplishment and mastery in learning) or a combination of both extrinsic and intrinsic desires.

References

Cerasoli, C. P., Nicklin, J. M., & Ford, M. T. (2014). Intrinsic motivation and extrinsic incentives jointly predict performance: A 40-year meta-analysis. *Psychological Bulletin, 140,* 980–1008.

Covington, M. V. (2000). Goal theory, motivation, and school achievement: An integrative review. *Annual Review of Psychology, 51*, 171–200. https://doi.org/10.1146/annurev.psych.51.1.171.

d'Ailly, H. (2004). The role of choice in children's learning: A distinctive cultural and gender difference in efficacy, interest, and effort. *Canadian Journal of Behavioural Science/Revue canadienne des sciences du comportement, 36*(1), 17–29. https://doi.org/10.1037/h0087212.

Eccles, J. S., & Wigfield, A. (2002). Motivational beliefs, values, and goals. *Annual Review of Psychology, 53*(1), 109–132.

Eccles, J. S., Wigfield, A., & Schiefele, U. (1998). Motivation to Succeed. In W. Damon (Series Ed.) & N. Eisenberg (Vol. Ed.), *Handbook of child psychology: Vol. 3. Social, emotional, and personality development* (5th ed., pp. 1017–1095). New York, NY: Wiley.

Feinstein, N. (2011). Salvaging science literacy. *Science Education, 95,* 168–195.

Gardner, R. C. (1983). Learning another language: A true social psychological experiment. *Journal of Language and Social Psychology, 2,* 219–239.

Gardner, R. C., Lalonde, R. N., Moorcroft, R., & Evers, F. T. (1985). Second language attrition: The role of motivation and use. *Journal of Language and Social Psychology, 6,* 29–47.

Gardner, R. C., & Lambert, W. E. (1959). Motivational variables in second-language acquisition. *Canadian Journal of Psychology, 13,* 266–272.

Gardner, R. C., & Smythe, P. C. (1975). Integrating research areas: A framework for second language studies. *Applied Linguistics, 9,* 198–217.

Glass, G. V. (1972). The wisdom of scientific inquiry on education. *Journal of Research in Science Teaching, 9,* 3–18.

Gottfried, A. E., Fleming, J. S., & Gottfried, A. W. (2001). Continuity of academic intrinsic motivation from childhood through late adolescence: A longitudinal study. *Journal of Educational Psychology, 93*(1), 3–13.

Hau, K. T., & Ho. I. (2010). Chinese students' motivation and achievement. In M. H. Bond (Ed.), *Oxford handbook of Chinese psychology* (pp. 188–204). Oxford, UK: Oxford University Press. https://doi.org/10.1093/oxfordhb/9780199541850.013.0014.

Hidi, S., & Harackiewicz, J. M. (2000). Motivating the academically unmotivated: A critical issue for the 21st century. *Review of Educational Research, 70,* 151–179. https://doi.org/10.3102/00346543070002151.

Hudson, G. (2000). *Essential introductory linguistics*. Oxford: Blackwell Publishers.

Kang, W. M., & Kang, W. Y. (Eds.). (2015). *Who says engineering is a fluffy discipline?*. Hong Kong: New Page Publishing.

Leung, K. S. (2002). Behind the high achievement of East Asian students. *Educational Research and Evaluation, 8,* 87–108.

Lukmani, M. (1972). Motivation to learn a second language and second language proficiency. *Language Learning, 7,* 261–273.

Music, G. (2014). *The good life: Wellbeing and the new science of altruism, selfishness and immorality*. New York, NY: Routledge.

Organization for Economic Cooperation and Development [OECD]. (2013). *Technical report of the survey of adult skills (PIAAC).* Retrieved from https://www.oecd.org/skills/piaac/_Technical%20Report_17OCT13.pdf.

Organization for Economic Cooperation and Development [OECD]. (2017). *PISA 2015 results (Volume III): Students' well-being*. Paris: Author. https://doi.org/10.1787/9789264273856-en.

Organization for Economic Cooperation and Development [OECD]. (2019, January). How is students' motivation related to their performance and anxiety? *PISA in Focus #92*. Paris: Author. https://doi.org/10.1787/22260919.

Perez, A., & Cromley, J. (2009, April). *Predicting science achievement from science media use and science enjoyment across 57 countries using PISA 2006*. Paper presented at the 2009 annual meeting of the American Educational Research Association, San Diego, CA.

Pintrich, P. R., & Schunk, D. H. (2002). *Motivation in education: Theory, research, and applications*. Upper Saddle River, NJ: Pearson/Merrill Prentice Hall.

Renninger, K. A., Nieswadt, M., & Hidi, S. (Eds.). (2015). *Interest in mathematics and science learning*. Washington DC: American Educational Research Association.

Ryan, R. M., & Deci, E. L. (2000a). Intrinsic and extrinsic motivations: Classic definitions and new directions. *Contemporary Educational Psychology, 25*(1), 54–67.

Ryan, R. M., & Deci, E. L. (2000b). Self-determination theory and the facilitation of intrinsic motivation, social development, and well-being. *American Psychologist, 55*, 68–78. https://doi.org/10.1037/0003-066X.55.1.68.

Salili, F., Chiu, C., & Lai, S. (2001). The influence of culture and context on students' motivational orientation and performance. In F. Salili, C. Y. Chiu, & Y. Y. Hong (Eds.), *Student motivation: The culture and context of learning* (pp. 221–247). Boston, MA: Springer.

Sansone, C., & Harackiewicz, J. M. (2000). *Intrinsic and extrinsic motivation: The search for optimal motivation and performance*. Cambridge, MA: Academic Press.

Svencer, B. (2012). *EDUtainment: Entertainment in the K-12 classroom*. Seattle, WA: Amazon Digital Services.

Taylor, G., Jungert, T., Mageau, G. A., Schattke, K., Dedic, H., Rosenfield, S., et al. (2014). A self-determination theory approach to predicting school achievement over time: The unique role of intrinsic motivation. *Contemporary Educational Psychology, 39*, 342–358.

Yu, C. H. (2012). Examining the relationships among academic self-concept, instrumental motivation, and TIMSS 2007 science scores: A cross-cultural comparison of five East Asian countries/regions and the United States. *Educational Research and Evaluation, 18*, 713–731. https://doi.org/10.1080/13803611.2012.718511.

Yu, C. H. & Wu, S. F. (2014, October). *Teaching math and science as a value system: Lessons from Asia*. Paper presented at Asian Conference on Education, Osaka, Japan.

Chapter 7
Time to Learn from Others

America is frequently characterized by its perpetual innovation and willingness to respond to challenges. Very often the US manages to reinvent itself within a short period of time. For instance, in 1957 American was shocked when the USSR launched the world's first human-made satellite named Sputnik (its literal meaning is "fellow traveler"). Before Sputnik, the US had already planned to launch its own Vanguard satellite, and Americans firmly believed that they would win the space race. In spite of this surprising initial humiliation, America took this opportunity to reorganize and reevaluate its educational system and research infrastructure, and also formed a close partnership between the federal government and university research (Boyle, 2008). Although before Sputnik the US scientific community and mass media had been pressuring for improvements in science and math education for a long time, the Sputnik challenge was a major catalyst for increasing the pace of the reform (Terzian, 2008; Urban, 2010). In general, Americans took the Sputnik challenge seriously. *Life Magazine* urged an education reform by publishing an article with this eye-catching headline: "Educators: Pleas for Better Math and US Billions" (as cited in Boyle, 2008, p. 375). In addition, the chancellor of the University of Kansas also asked for education reform in response to the Sputnik challenge, saying, "We still consider education a luxury rather than a necessity. The little satellite says that our schools are no longer a luxury but are as important as the food we eat, the cars in which we ride, the clothes we buy" (as cited in Boyle, 2008, p. 375). In 1958, US Congress passed the *National Defense Education Act* (NDEA), a four-year program that provided 40,000 loans, 40,000 scholarships, and 1,500 graduate fellowships to encourage students to study science, math, foreign languages, and other fields, as well as to provide funds for schools to acquire the necessary resources for improving education. This act was considered unprecedented and revolutionary because it broke the tradition of a hundred years of federal inaction in American education. In the year 1953 alone, the US government spent $153 million on education, $10 million of which went to colleges and universities. By 1960, the combined funding grew almost six-fold (Urban, 2010). This act also laid the foundation for many other subsequent acts in education, including the *Elementary and Secondary School Act* of 1965 and the *Vocation Education Act* of 1963 (Boyle, 2008). In accessing the long-term effect

C. H. Yu and H. S. Lee, *Creating Change to Improve Science and Mathematics Education*, https://doi.org/10.1007/978-981-15-3156-9_7

of the Sputnik challenge and the NDEA, Boyle (2008) claims that many children "probably would not have been able to attend college at all without the existence of the NDEA" (p. 383). Although the Apollo Program is not directly related to NDEA, the national spirit pushing for better scientific research eventually put several American astronauts on the moon in 1969. The rest is history. This example shows that a paradigm shift in education is indeed within our reach.

The United States once was open to adopting effective and promising educational ideas from foreign countries, mostly European countries, such as Germany (Tucker, 2011). However, in the aftermath of the Cold War, as the US has emerged as the sole superpower nation on earth and experienced an exponential growth in its economy and technology, some from the US have grown to believe that "we have little to learn from anyone" (Tucker, 2011, Chap. 6, p. 2). The lagging academic performance of US students and high attrition rates among STEM pursuers, compared with the fact that other countries are outperforming the US by far, is calling this old belief into question. The US must confront the problem that has been lurking around for decades. Several US teachers shared their perspectives toward the current US education system. One of them articulated the minds of the authors in a metaphorical but extremely clear fashion:

> Some say that this change has been a long time coming. There is an analogy that uses fairy tale character Rip van Winkle to describe this; Rip van Winkle has just woken up from his hundred-year slumber and stares in amazement about how much everything has changed in the time that he was asleep. He almost did not recognize anything, until he went into a classroom. Rip van Winkle recognized immediately that it was a classroom because nothing much has changed in the K-12 educational system since he fell asleep in 1906.

The current state of US education calls for a humble mindset to admit that the time has come to look for answers from other countries in the world. We argue that Hong Kong should be considered as a worthwhile model from which the US can learn. The main argument is the remarkable cultural compatibility argument of the Hong Kong education system, which stands out from East Asian countries. More importantly, abilities are not predetermined, and therefore each and every system is capable of transformation. In their study, Geary, Salthous, Chen, and Fan (1994) found that younger Chinese adults (average age = 19.4) outperformed their American counterparts (average age = 20.1) and older Chinese adults (average age = 65.6) in math, but the performance of older Chinese adults (aged 60–77) and that of American adults (aged 57–85) were comparable in all ability measures, implying that the advantage of Chinese adults over American adults in complex math is a relatively recent phenomenon. Similarly, the younger generation of Singaporeans (aged 16–24) outperformed the older generation (aged 55–64) in 2016 PIAAC (OECD, 2016). One plausible explanation is that the younger cohort has had more frequent exposure to better education, while the older cohort has been away from the academic environment for decades (or they were taught by traditional, ineffective methods). These findings are encouraging to students, educators, policy makers, and other stakeholders in education. Perhaps good pedagogical strategies and education policy could make a big difference for the next generation, but it requires openness and courage to break with the status quo.

7.1 Looking for Answers Within the International Community

Tucker (2011) argues that the policies directing the education system are what senior US policy makers should look at thoroughly. In any system, the reformation of an existing framework presents worries and challenges. Therefore, we tend to look for adjacent examples, such as Canada or other European countries that may share some cultural similarities. This approach is undoubtedly reasonable. To see changes produce an effective outcome in the short term, targeting culturally similar countries or regions as models serve the purpose. Nonetheless, the authors suggest that breaking the old mold is necessary. We need to look further. Not only that, we also need to try harder for a longer time to make educational innovations take effect. It will take rigor to solve the problem. It will take a long-term commitment to complete the task.

7.1.1 How Europeans Transformed US Science

To add more support for this argument, the West learning from the East (and vice versa) is not a new idea. Successfully transplanting a foreign model recurs in history all the time; indeed, it has happened in US history before. Interestingly, the US has tremendously benefited from two of its former enemies. Before World War II, Europe was the center of research in advanced physics, housing prominent scientists such as Niels Bohr, Werner Heisenberg, Enrico Fermi, Paul Dirac, Albert Einstein, and many others. Before the rise of the Nazi Party in 1933, German scientists had won 30% of all Nobel prizes in science since the first award was conferred in 1901 (Zuckerman, 1977). Due to the research gap, American graduate students in physics had to study in Germany in order to learn cutting-edge physics. For instance, future Nobel laureate Isidor Rabi was a mentee of German physicist Otto Stern; likewise, the future head of the Manhattan Project, Robert Oppenheimer, was a student of Max Born (Nachmansohn, 1979). During and after World War II, many European scientists immigrated to the US. Specifically, between 1930 and 1941, twelve European Nobel prize winners in science came to the United States to escape Nazi and Fascist persecution. Not only did these immigrants contribute to the invention of the atomic bomb, leading to the Allies victory in World War II, but they also substantively transformed science education and scientific research in the US by introducing innovative ideas and rigorous research methods, as well as building a mentorship model. After World War II, European scientists who stayed in the US mentored many future American Nobel laureates and also built a mentoring network for subsequent Nobel Prize winners and renowned scientists (Chariker, Zhang, Pani, & Rouchka, 2017; Zuckerman, 1977).

7.1.2 How Japan Influenced US Management and Technology

In addition, between the late 1970s and early 1990s, US corporations faced strong challenges from Japan. American companies were driven out of the home electronics market due to the superior quality and competitive pricing of Japanese products, while the US microchip and auto industries were also severely wounded (Grove, 2010). In spite of this alarming situation, some Americans still refused to learn from Japan by invoking the dogma of cultural difference (Vogel, 1999). Nonetheless, a few US corporations, including GM, GE, and Boeing, admitted their shortcomings and humbly imported the Japanese model (Holden, 1986). Specifically, Boeing sent hundreds of its executives to study in Japan and also hired retired Toyota people as consultants (Bradford, 2006). Furthermore, after spending 11 years working for Toyota Motor Corp. and learning from its lean management systems, in 1983 John Shook facilitated the formation of a joint venture between GM and Toyota in Fremont, CA, which was called New United Motor Manufacturing, Inc. (NUMMI). By implementing the Toyota system, absenteeism of GM autoworkers in Fremont dropped from over 20% to 2%, and the plant became a role model in the auto industry (Shook, 2010). According to Shook, had US automakers adopted the Japanese methods earlier, their financial hardship would have been much less severe. In the era of globalization, the line between cultures is blurred. When GM adopted the Toyota model, in some sense GM was relearning the American quality control model, which was introduced to Toyota by American consultant Edward Demining in the 1950s (Boardman, 1994).

7.1.3 The UK Learned from Shanghai in Math Education

In fact, East and West have also exchanged ideas about education policies quite frequently. When Shanghai students shocked the rest of the world with surprisingly high performances in math and science, the Western world began to study Shanghai's teaching model (Tan, 2013). The United Kingdom came to the forefront and decided to adopt the Shanghai math curriculum. To some, this decision seemed a bit reckless and desperate. Yet, the UK had been performing at poor levels in terms of international tests. UK pupils lagged behind in the 26th position, whereas Shanghai has consistently scored the top in PISA. The problem was not one of slow or stagnant growth, but of rapid decline compared with its global rivals, mostly in Eastern countries. Thus, the UK's decision to import Shanghai's method was indeed out of desperation, to cease falling further behind. However, as will be discussed later, the UK's decision was not a hasty one. Before looking into the success story, however, the authors feel the need to address an obstacle toward importing a foreign education practice: the cultural barrier.

7.2 Cultural Barrier

The authors are fully aware of the fact that a direct comparison between the US and Hong Kong may appear implausible. To begin with, the US is a relatively young and large country comprised of multiple states. Each state demonstrates variations in culture, history, and belief, while together they share some commonalities as America, one nation. Heterogeneous racial/ethnic groups comprise the US population, to which increasing numbers of immigrants from different nations of the world are added every year. In contrast, Hong Kong is a small region with thousands of years of rich history and culture that coherently bonds the people of Hong Kong together. Hong Kong shows a fairly high degree of homogeneity in its population. These fundamental differences might cast doubt about the viability of adopting Hong Kong's approach.

Within the sample of our study, this concern was shared by educators in both the United States and Hong Kong. A Hong Kong-born professor interviewed by us also argued, "Comparing HK against the US is comparing a city to a country. There are too many inherent differences to make this comparison fruitful". One US high school teacher commented, "Learning from another country/culture can easily bring confusion if not conflict when the policies are replanted without thorough research taking place before the implementation". The argument of cultural differences between Hong Kong and the US represents the main viewpoint of many scholars and educators who have posed serious doubts on the compatibility of Eastern Asian education models with US students. Scholars insist that borrowing a foreign model needs to factor in several variables, such as context, causality, and transferability (Cowen, 2006; Phillips, 2006). The issue of context is based on the idea that no model is one-size-fits all. For instance, certain philosophies are more likely to fit certain cultural contexts than others. It becomes an issue of reliability: although the policy makers ensure the reproduction of the model, it may not be replicated in a different environment. Furthermore, prior to adopting an education system from a foreign country, one must inspect the causal relationship between a certain education system and the academic performance of students in a particular subject. Lastly, the implementation of the desirable education model benefits from the ease of transferability. For this reason, Western countries tend to learn from neighboring nations or cities, which already have a similar education model. Experts and educators alike fear that when the criteria based on the aforementioned critical factors are not met, the likelihood of successfully importing a foreign model stands little chance.

Others who support the idea of the West learning from the East tell a different story. One of the most popular arguments is that East Asian countries have themselves adopted educational practices modeled after Western ones. As discussed in Chap. 2, the UK played a significant role in developing and reforming Hong Kong education practices. Interestingly, the idea of teacher–student interaction originated from the West, but in their field observation, Miao and Reynolds (2014) found that in the Chinese classrooms whole-class interactive teaching was used 72% of the time, compared to only 24% in England. Some core ideas for the contemporary education system and higher learning institutions in East Asian countries were first developed in and imported from the Western world since the 1800s. Education models and university ideas of the United States, the United Kingdom, Germany, and France were adopted and reconstructed (Shin, 2012). Scholars and educators cited several East Asian nations, including Japan and Korea, which adopted or adapted Western educational models (Shibata, 2005; Shin, 2012). Since its early encounter with the Western world in the mid-1800s, Japan has shown the desire to adopt the education model of the United States (Shibata, 2005). Such attempts seemingly paid off as Japan experienced relatively rapid growth in its economy and technology in comparison to neighboring Asian countries, such as China and Korea. Even after World War II, Japan continued to aggressively adopt the education model of the United States (Shin, 2012). South Korea, where Western missionaries visited and founded Western-inspired schools, has also grown to show an exceptional openness to Western ideas around universities. To this day, Korean universities owe several major ideas and structures around academic courses, teaching style, and research methods to Western universities. Furthermore, many of the most prestigious universities in South Korea were founded by Western missionaries or Japanese colonial governments, which were themselves heavily influenced by the Western world. These aforementioned East Asian countries, among others, have demonstrated an appreciable willingness to embrace some Western educational ideas that they found effective and beneficial.

Certainly, change should occur with caution. Duffy (2012) found that South Koreans paid a high price for their superior academic performance, evidenced in part by their high suicide rate; thus he warned that the US should not blindly adopt the Korean approach. Jeynes (2008) took into account the arguments of two opposing parties and suggested taking the "middle road" (p. 900). The two possibilities for ways that the US might learn from East Asian education in spite of projected cultural differences are: (1) Selecting applicable qualities that are shared by multiple Eastern Asian education systems that have proven to be effective. This criterion gives the learning country some degree of confidence that there is a positive causal effect of a certain education system on students' academic outcomes. (2) Targeting and reimporting education paradigms that were originally borrowed from the United States and have proven to be successful in East Asian contexts. In this way, the learning country can minimize the unpredictability of the context-based compatibility, as well as maximize the transferability of the model. If the models that once were core beliefs in the United States have proven their functionalities in the East Asian country, it is safe to assume that readopting the old model or the new integrated model may yield similar results. The authors, who also relate to both sides of the debate, fully agree with the aforementioned qualities of the education model candidates. Given this, we strongly argue that Hong Kong is an appropriate model for the US, as we firmly believe that Hong Kong is the *middle road* to success that meets the aforementioned criteria.

Last, the authors are well aware of certain major cultural differences due to our experiences with both Asian and Western cultures. Hence, we propose small-scale reforms instead of a radical revolution. As mentioned in Chap. 1, this book has addressed malleable factors only and avoided going into the debate of parenting styles and other factors that are difficult, if not impossible, to change. Rather, the focus of this book has been on certain factors that transcend cultures. For example, psychological research suggested that there might be a universal structure in problem-solving (Hooker, 2017). If this is the case, it would be difficult to make a counterargument that the method of promoting problem-solving and critical thinking in Hong Kong cannot work in the US. In a similar vein, Robinson and Aronica (2015) argued that children are naturally curious; therefore, given proper instruction, they can be guided to be creative. Again, there is no reason to believe that American students are less curious than their Hong Kong peers.

7.3 The Best of Both Worlds: Where East and West Meet

The great advantage of Hong Kong as an ideal learning model is the uniqueness of Hong Kong in terms of its blended Western and Eastern culture. Hong Kong is known for its collaboration between Chinese and Western traditions, largely due to its history. Hong Kong was formerly one of many colonies of the British Empire for more than 150 years, from 1841 to 1997. During this time, Chinese people in Hong Kong were substantially influenced by Western educational ideas in various aspects, including the education system (Tsui, 2007). Frederick Stewart is called "the Founder of Hong Kong Education" because he introduced a Western-style education model to Hong Kong when he was the headmaster of the first government school in 1862 (Bickley, 1997). Hong Kong education today is still largely modeled on the English system, but it is also highly influenced by the Confucian heritage, which embraces the idea of "self-cultivation", a learning process through continuous self-examinations and corrective actions. Thus, the Hong Kong model is a fusion of the East and the West (Luk-Fong, 2006).

To elaborate on the Confucian value of scholarship, Hong Kong people greatly value an ethic of rigorous work and persistence in search of knowledge and are eager to attain high performance as its fruit (Lam, Ho, & Wong, 2002). One of the core perspectives the Confucian tradition acknowledges is the significance of education: it is not only perceived as a matter of personal improvement but also of societal and national progress (Lee, 1996; Tu, 1993, 1996). Another belief is the significance of one's continuous growth in education, as education is perceived as the path to reach the fullest potential of the person and the society (Oh, 1991). In this continuous journey of learning, willingness and effort are considered the most pertinent factors or qualities to which one must dedicate oneself (Tweed & Lehman, 2002). Lastly, Confucian philosophy advocates for the potential of all people, where everyone is eligible to be educated to become a better person with greater knowledge and a disciplined mind (Tan, 2013). The conviction of educability and perfectibility strongly argues that all children, irrespective of their genetic and/or socioeconomic factors, can obtain success in any field with good teachers and hard work.

In Hong Kong, many students, parents, teachers, professors, and government officials are united in one mind. They recognize the importance of quality education, the need to provide it to all students from all backgrounds, and the expectation of satisfactory learning outcomes. Parents, who are fully aware of the fast-moving job market, make themselves busy finding ways and resources to raise their children as competitive runners in this race. Teachers and professors are ready to capitalize on the advantage of students' early and consistent exposure to a rigorous academic environment. They also strongly hold the perspective that it is never too late for one to pick up where one left off and that one should never give up on learning. Professors are not fearful of challenging students with coursework of advanced qualities that demand students' rigorous efforts. Rather, they enjoy it, believing that by doing so, students are more equipped not only to survive but to rise above their competitors in the world. Students, who are driven both intrinsically and extrinsically, are reinforced to push themselves. The Hong Kong government, which has been continually taking a keen interest in improving its education system with a focus on STEM, has attempted to ensure that it concretely plans for educational reformation to be delivered to local schools and colleges, as explained in Chap. 2. These seemingly distinct parties of students, parents, educators, and government officials—which comprise the educational system of Hong Kong—work together with a shared vision of and value for quality education, the essential need for learning, and the power of knowledge.

As for Western values, Hong Kong has attentively welcomed the student-focused approach from the Western world. People in Hong Kong were once greatly influenced by Western power and culture, unlike the rest of mainland China. Although over 95% of its population is of Chinese origin and is rooted in the Confucian tradition, Hong Kong exhibits an exceptionally westernized view of society in comparison to China and other Asian countries that have been heavily influenced by Confucian belief (Lam et al. 2002). Its highly westernized thinking is most remarkably embedded in the education system. One prominent idea is the academic atmosphere, which spurs students' active communication of their ideas between peers and teachers. For instance, in mainland China, students are encouraged to obey and follow the teacher and authorities. This atmosphere may stifle the development and communication of several attributes, such as creativity. However, in Hong Kong, educators encourage more diversity of student opinions on the basis of promoting academic freedom. This approach is best known as the Socratic Method of learning, which takes the form of dialogue between the learner and the instructor (Tweed & Lehman, 2002). Socrates showed his tendency to evaluate the already existing theories and beliefs. He supported the habit of questioning other scholars and encouraged it among his students such as Plato as well, as he believed that through repetitive logical doubt the knowledge of a person and humanity as a whole would continue to progress.

In the eyes of traditional Confucian believers, students' practices of repeatedly and overtly questioning teachers in public may come across as defiance of authority (Lee, 1996). In Chinese culture, respect for authority figures (such as parents, elders, and teachers) is considered an ethical pillar. If students have doubts or disagreement with their instructors, they are taught either to accept them quietly or to delay their question. Given this factor, Hong Kong educators strive to create an academic setting where both teachers and students can have an open dialogue as their Western counterparts are doing (Tsui, 2007). In this new setting, students are not afraid of addressing their doubts and questions in the classroom, yet they are to ask them in a respectful and humble manner. Teachers encourage students to share their questions both in private and in public, and they are trained not to feel intimidated or disrespected by receiving some challenging questions. Instead, they try to widen the learning space for students and give them the freedom to independently reevaluate, question, and recreate the knowledge they have acquired from their teachers. Also, the schools utilize creative and various methods of teaching to increase student engagement. Classes are now expected to be multimodal; teachers are expected to upgrade their methods of teaching with creative, up-to-date resources; and students are allowed to incorporate a variety of methods to present their ideas.

Lastly, Hong Kong government and schools take students' needs, personal interests, learning styles, and talents into account. In short, educators are willing to customize the teaching curriculum, teaching styles, and assessment materials for their students in order to maximize their learning experiences. This learner-focused approach, which demonstrates a core educational value in the Western world, has been successfully adopted into the Hong Kong education system. Below are some of the examples provided by Hong Kong teachers in secondary schools that demonstrate Western styles of teaching:

- "Open-mindedness, student-centeredness, setting different goals for different students."
- "Various modes of activities depending on time and ability of students. Plus alternative suggestions in different topics' learning for students with different ability."
- "Provide different rigorous academic training to different abilities of my students."
- "I use apps to let students upload their exercises rather than asking them to write their answers on the board. Some students are too shy to share their answers. I look at what they submitted and select some of them to discuss. I also ask students to discuss the exercises in team. It is better for students to teach each other. Students might not understand the jargon used by the teacher, but they share the same 'language.'"
- "Create joyful atmosphere. Allow educational games for learning. Welcome questions. Welcome suitable jokes from students. Let students express their learning styles."

One professor who currently teaches at a university in Hong Kong concisely expressed how well-mixed Hong Kong education's system is, in terms of integrating Western ideas (e.g., student-centeredness and interactive teaching styles) and Eastern ideas (e.g., respect for the teacher, persistence in improvement):

> Chinese students are respectful to their teachers and the learning. Hong Kong teachers, who are in fact regarded as prestigious professions, put their efforts to constantly seek change so that teachers themselves are not left behind in terms of the quality of education they provide. Most schools are active in looking for innovative and interactive teaching methods because they want to be more attractive to students and also to train their students to outperform others. Hong Kong Government has various funding for innovative STEM education to support schools, teachers, and students.

Given this, the authors perceive Hong Kong as a successfully blended entity of both East and West, the best of both worlds. When it comes to its education, Hong Kong has ventured to grasp the best ideas in the West in the context of its Confucian foundation. Hong Kong aggressively seeks for the betterment of its educational system outside of its root culture and traditional models. Policy makers and educators have shown passion and readiness to adopt and adapt the best practices. It is apparent that their efforts have produced some rich fruit. In some sense, importing Hong Kong educational model is to reimport an Asianized Western model.

7.4 Change with Caution

Although the culture of Hong Kong may resemble that of the US more than most of Eastern Asian countries, there is no doubt in our minds that a certain degree of cultural differences between the two countries remains, which may lead to currently unforeseeable issues. Taking these into account, the authors believe that the Hong Kong model should be imported and implemented, starting with small target groups. Although our claim seems bold, we suggest that it be put into effect thoughtfully. The principals and educators of secondary schools will potentially be the main executors for enacting the Hong Kong model. As addressed in Chap. 1, Asian immigrants brought their *success frame* from their country of origin and successfully reconstructed a well-operational academic system in the US. One teacher who currently works at a private academy located in the heart of Los Angeles informed us that:

> Although our school has mostly Asian kids, we do have a couple of non-Asian students in each class year. Teachers may pay some extra care to those non-Asian students, in terms of their peer relationship and maybe (accounting) the cultural difference between them and the Asian kids, but that's about it. We hold the same expectation for them as we do for Asian students. We do not excuse them for anything when it comes to us teaching and them learning. Instead, we make sure all students experience learning in their own creative ways. Together, they go through the same core classes, college prep, same monthly check-up. Individually, teachers pay attention to each kid's personal interest and preferences. We also make sure their parents know how their kids are doing at school, we give them monthly reports about student's performances, sometimes weekly if needed to have everyone on the same page… We just believe that every kid has the potential to be great. We know that because, every year, most of our students, including the non-Asian students, show same great quality of performance and admission rates to their so-called dream schools. I am personally very proud of them. It's always good to see a kid growing and as a teacher I see many of them doing so. We (teachers) just want them to excel. We want them to be the leaders. We want them to live their dreams. That's all.

This private academy was founded by a first-generation Asian immigrant who had a passion to teach not only the children of his own people group but everyone in the area. He wanted to open doors to all who aspired to leverage themselves to grasp better and greater opportunities in their lives through educational excellence. That passion seems to be slowly emerging into the light. More recently, the school, which has predominantly consisted of Asians from various social levels, is now gaining the attention of students and parents of other races and ethnicities and experiencing more diversity in the student body. All students, regardless of their race and socioeconomic backgrounds, receive the same level and degree of education that demands a similar intensity of effort. The teachers at this school are also racially and culturally diverse, but they commit to the same mentality and method of teaching employed by the school principal. In several ways, the school's teaching model greatly resembles that of Hong Kong education. Teachers and parents are trained to pay attention to students' academic development and achievement. This example paints a hopeful future. The education model seems to confirm one of its core values: everyone can learn and excel regardless of their social class, culture, and race.

7.5 One Small Step, One Giant Leap

The authors believe that the safest and quickest way to improve weaknesses is to watch and learn from the best. As the United Kingdom's attempt to import Shanghai's teaching model continues to unfold, it is notable that the UK initiated this potentially massive educational reformation project with a series of small programs. One of the program initiatives was a Shanghai–England teacher exchange (Low, 2017; Weale, 2015). The United Kingdom Department for Education invited about seventy teachers from Shanghai. The teachers are expected to teach in local secondary schools with teaching philosophies and strategies endorsed in Shanghai. The main objective of Shanghai teaching model that the United Kingdom took great notice of is the so-called "mastery method". The teacher pays a concentrated effort on content and delivers concepts thoroughly in a step-by-step process. For instance, teachers cover a single mathematical concept with all students. Only after every single student in the classroom is clear on the subject does the class moves forward. It is a step-by-step mastery method. Then, the performances of students will be evaluated in comparison to their counterparts elsewhere in England.

Given the compelling evidence of academic excellence in Shanghai, Nick Gibb, school minister in charge of the Shanghai project, believes that this exchange program is worthy of trying. In fact, he perceived that this is "one of the most valuable education initiatives undertaken by our government over the past few years" (Gibb, 2015, p. 1). To the skeptics, he responded:

> There are many who try to attribute Shanghai's achievement to supposed socio-cultural explanations. These critics point towards "tiger mothers", China's growing economy, or the deferential nature of Chinese society to explain away their success. These crude national stereotypes are not the issue; it is simply narrow-mindedness to insist that there is nothing we can learn from finding out what happens inside Shanghai's classrooms (Gibb, 2015, p. 3).

Among many who support this movement, David Reynolds, a professor of education at Southampton University, projected that UK students may exhibit improved math scores in the 2018 PISA, after having benefited from the Shanghai teaching method (as cited in Weale, 2015, p. 2).

The United Kingdom, a nation that has been one of the most influential leaders of the world, humbly admitted that its educational system needs improvement. And the UK wants to learn from the best. In the eyes of the United Kingdom, the best model in Shanghai, a city that has consistently been ranked as the top education system over many rounds of the OCED's PISA tests (Coughlan, 2013). The authors believe that the most appropriate model for the United States is Hong Kong, a system that creatively integrates the elements of East and West.

7.6 Conclusion

In the previous chapters, the authors have introduced the overarching theme, core principles that could serve as policies, and examples of effective practices both inside and outside of the classroom that accounts for the success of Hong Kong's education system. The overarching theme of the current Hong Kong education system is *Learning to Learn*, which encourages people's autonomous learning mentalities and skills throughout their entire lifespan. Under this umbrella rests core learning qualities: creativity, problem-solving, self-efficacy and productive failure, and motivation.

1. *Learning to learn* (the overarching theme): Learning to learn is essentially lifelong learning. When the goal of the student is nothing more than earning a good grade, it is not surprising to see a learner forget 90% of the content after the exam. If retaining old information remains a problem, so will be the ongoing learning of new information. The Hong Kong education framework attempts to rectify this situation by moving its emphasis away from the exam-oriented culture to the value of learning, especially lifelong learning. Simply put, learning to learn is based on the philosophy that the learning process is an active construction of knowledge by the learner. If the learner only passively absorbs information from the instructor without taking ownership of his learning experience, even a so-called "good" student will cease to grow after graduation. However, today the global economy has transformed from an industry-centric mode to an information-centric mode, meaning that the state-of-the-art today might become outdated a year later. To cope with the new demands of global trends, it is mandatory to redefine education through the lenses of lifelong learning and learning to learn. By doing so, Hong Kong teachers appeal to the human sense of wonder and the mysteries of the universe. It is a lifelong endeavor to pursue inquiry and attain intellectual fulfillment.
2. *Creativity, work ethic, and foundational knowledge*: There is a growing consensus that education should aim at cultivating creativity so that students can succeed in modern economies that emphasize new ideas and innovation. However, creativity cannot be accomplished by treating it as an empty slogan. Creativity entails an inquisitive mentality, a robust imagination, and a commitment to critical reflection. It is the conviction of the authors that the education practice following *No Child Left Behind* tends to promote conformity and a "cookbook" approach that

fails to foster creativity. There is a popular belief that Asian countries/regions, including Hong Kong, emphasize rote learning. Yet Hong Kong students excel in PISA and IMO, which demand application and logical reasoning, not memorization. Indeed, many Hong Kong math and science teachers utilize pedagogical methods that aim to promote various aspects of creativity, including combination thinking (integrating seemingly unrelated events) and counterfactual reasoning (considering otherwise). Nonetheless, work ethic and basic knowledge are the prerequisites of creativity. Western scholars have found a strong association between creativity and certain personality traits (e.g., tolerance of ambiguity, preference for disorder, willingness to take risks, and persistence). Specifically, creativity emerges when the learner can challenge the status quo, stick with difficulty, consider alternatives, and dare to be different at the end. These psychological traits can be shaped by proper pedagogical methods, such as urging students to solve ill-defined problems.

3. *Problem-solving*: "If you've never failed, you haven't tried hard enough". Failure is the precursor of success. However, many students cannot "fail well" because they are overprotected by highly structured learning environments and abundant learning resources. The premise of problem-solving is that real-life problems are inherently ill-defined, with relevant information missing and limited resources. Hence, in the view of Hong Kong educators, giving detailed, step-by-step guidelines does a disservice to the learner. Hong Kong's education program has designed a school curriculum to instill a problem-solving mindset, to present consistently challenging, creative, and advanced problems, and to train students to dedicate their rigorous effort throughout the process of problem-solving. It is the firm belief of Hong Kong educators that problem-solving is both the means and the goal of math and science. By exposing students to extraordinary and complex problems at an early age, teachers test the concreteness of students' understanding and mastery of their subjects, the level of their critical/analytic thinking skills, and their determination to solve a problem. Students also expect to confront a variety of problems in math and science that mandate a degree of struggle, yet they are willing to accept those challenges. By doing so, they can develop a domain-general problem-solving competence that can be widely applied to different contexts and settings. To recap what was discussed in Chap. 2, the problem-solving skills acquired from math and science learning, such as observing, imaging, abstracting, patterning, and empathizing, can be transferred to art, music, and other nonscience domains.
4. *Self-efficacy and productive failure*: Although prior research suggested a clear positive link between how we feel about ourselves and our overall wellbeing, recent international assessment data (e.g., PISA, PIAAC, and TIMSS) has revealed that the relationship between self-efficacy/self-concept and academic performance is nonlinear. When the ego of the learner is improperly inflated, it hinders them from further learning. The reason is simple: When a person has an inflated ego, he overlooks the fact that there is always something that he does not know, and as a result he ceases to absorb new information or see things from a new perspective. This attitude is also counterproductive in collaboration, which

requires give-and-take between team members. But an egoistic person does not realize that someone around him may know more. As a remedy, Hong Kong educators have introduced the approach of productive failure in order to expose the knowledge gap of the learner, leading them to build a modest character, while also creating teachable moments. To be specific, the learner can improve their skill at a much faster pace by making an accurate assessment of their ability and realizing their misconceptions.

5. *Motivation*: It is the conviction of the authors that the demarcation of intrinsic and extrinsic motivation is a false dichotomy. In most situations, people carry their actions out of a combination of both types of motivation. Additionally, we proposed that it is an urban myth that Hong Kong students learn math and science solely for practical and exam-oriented purposes. To reiterate, Hong Kong is not a high-tech economy. Hong Kong lacks research institutes like JPL or deep-pocketed funding agencies for scientific research, like NSF and NIH. Nonetheless, Hong Kong students are willing to learn math and science, driven by a mixture of personal passion and a variety of external rewards. This atmosphere has successfully been cultivated by Hong Kong's educational framework. To be specific, the Hong Kong system does not aim to make every learner a scientist or an engineer, nor does it design the curriculum solely in response to the specific labor force needed by the economy. Rather, the curriculum aims to bring lifelong benefits to the students. Scientific literacy is desired by everyone because the government and the people believe that when everyone is eligible to be a scientifically educated citizen, the overall extent of personal and national wellbeing improves. To be specific, in a knowledge-based society, scientifically educated citizens can participate in public decisions, which requires an open dialogue between professional scientists and laypeople in a transparent and democratic process, such as driving environmental policies against climate change.

Despite the apparent cultural difference, it is crucial to point out that Hong Kong's approach, which emphasizes work ethic, motivation, and foundational knowledge, is readily transferrable to the US context. Prior research conducted in the US indicates that work habits (student effort to learn), motivation (student enthusiasm), and background information (adequacy of students' background and preparation for the course) are potent predictors of the rating of their learning experience and the effectiveness of the teacher. Classes consisting of students who possess the above attributes tend to receive higher course ratings, regardless of who teaches them (Benton & Li, 2017; Benton, Li, Brown, Guo, & Sullivan, 2015).

These qualities of Hong Kong education are malleable factors that the US can actively adopt and implement for its purposes. With the paradigm of Hong Kong education, the trajectory of the current educational performance of the US will shift its direction toward uplifting success, not continuing downfall. Several methods of importing Hong Kong's education model exist. The authors strongly suggest we follow the UK's approach to Shanghai, importing little by little with close observation and follow-up evaluations. America can invite a small group of teachers from Hong Kong into classrooms in secondary schools. US teachers may visit Hong Kong to experience Hong Kong's educational environment and practices and bring its lessons home. With an open mindset and a willingness to embrace learning, we can come up with a variety of ways to benefit our education system.

Epilogue

Learning to learn may serve as a core value of the educational benchmarking strategy of the US. Learning never ends, and the search should never stop. In *Surpassing Shanghai*, Tucker argues that international benchmarking takes it fullest form when the countries compare broad goals, policies, practices, and institutional structures, as well as relative standing on common measures, in order to understand what another country is trying to achieve, how it has gone about achieving it, what it would have done differently, what mistakes it made and how it addressed them, which factors most account for its achievement, and so on (Tucker, 2011, Chap. 6, Sect. 4, p. 2).

The authors strongly believe that the US has much to gain by learning from Hong Kong's education system. Nonetheless, the expansion of our idea reaches out toward greater momentum for this nation. America should be on the continuous lookout for successful education models and socioculturally compatible policies and practices around the world. The US should be driven to leverage its educational resources and policies so that students not only survive but rise above their competitors. Only the ones who desire success will rightfully gain it.

References

Benton, S., & Li, D. (2017, July). *IDEA Research Note#6. Validity of the IDEA student ratings of instruction student characteristic items*. Manhattan, KS: The IDEA Center.

Benton, S. L., Li, D., Brown, R., Guo, M., & Sullivan, P. (2015). *IDEA Technical Report No. 18: Revising the IDEA student ratings of instruction system*. Manhattan, KS: The IDEA Center.

Bickley, G. (1997). *The Golden needle: The biography of Frederick Stewart (1836–1889)*. Hong Kong: David C. Lam Institute for East West Studies.

Boardman, T. J. (1994). The statistician who changed the world: W. Edwards Deming, 1900–1993. *American Statistician, 48*, 179–187.

Boyle, R. (2008). A red moon over the mall: The Sputnik panic and domestic America. *Journal of American Culture, 31*, 373–382. https://doi.org/10.1111/j.1542-734X.2008.00684.x.

Bradford, M. (2006). Boeing flies high the Toyota way. *Strategic Direction, 22*(8), 14–16. https://doi.org/10.1108/02580540610675399.

Chariker, J. H., Zhang, Y., Pani, J. R., & Rouchka, E. C. (2017). Identification of successful mentoring communities using network-based analysis of mentor–mentee relationships across Nobel laureates. *Scientometrics, 111*, 1733–1749.

Coughlan, S. (2013, December 3). Pisa tests: UK stagnates as Shanghai tops league table. *BBC*.

Cowen, R. (2006). Acting comparatively upon the educational world: Puzzles and possibilities. *Oxford Review of Education, 32*, 561–573.

Duffy, C. (2012, May 23). Korea's model education system? Think again. *Teachers College Record*. Retrieved from http://tcrecord.org.

Geary, D. C., Salthous, T. A., Chen, G. P., & Fan, L. (1994). Numerical cognition: Are US-East Asian ability differences a recent phenomenon? *Developmental Psychology, 32,* 254–262.

Gibb, N. (2015, November 26). The math teachers of Shanghai have the perfect formula for learning. *The Guardian*. Retrieved from https://www.theguardian.com/commentisfree/2015/nov/26/maths-teachers-shanghai-china-uk.

Grove, A. (2010). *Only the paranoid survive*. Danvers, MA: Crown Business.

Holden, C. (1986). New Toyota: GM plant is US model for Japanese management. *Science, 233,* 273–277.

Hooker, C. (2017). A proposed universal model of problem solving for design, science and cognate fields. *New Ideas in Psychology, 47,* 41–48.

Jeynes, W. (2008). What we should and should not learn from the Japanese and other East Asian education systems. *Educational Policy, 22,* 900–927.

Lam, C. C., Ho, E. S. C., & Wong, N. Y. (2002). Parents' beliefs and practices in education in Confucian heritage cultures: The Hong Kong case. *Journal of Southeast Asian Education, 3*(1), 99–114.

Lee, W. O. (1996). The cultural context for Chinese learners: Conceptions of learning in the Confucian tradition. In D.A. Watkins & J.B. Biggs, (Eds.), *The Chinese learner: Cultural, psychological and contextual influences* (pp. 25–41). Hong Kong: Comparative Education Research Centre and Victoria, Australia: The Australian Council for Educational Research.

Low, H. (2017, January 20). Should all countries use the Shanghai math method? *BBC*. Retrieved from https://www.bbc.com/news/magazine-38568538.

Luk-Fong, P. (2006). Hybridity in a guidance curriculum in Hong Kong. *International Journal for the Advancement of Counselling, 28,* 331–342.

Miao, Z. Z. & Reynolds, D. (2014). *How China teaches children maths so well*. Retrieved from https://www.researchgate.net/publication/299426602_How_China_teaches_children_maths_so_well.

Nachmansohn, D. (1979). *German-Jewish pioneers in science (1900–1933)*. New York, NY: Springer.

Oh, T. K. (1991). Understanding managerial values and behaviour among the gang of four: South Korea, Taiwan, Singapore and Hong Kong. *Journal of Management Development, 10*(2), 46–56.

Organization for Economic Cooperation and Development [OECD]. (2016). *Skills matter: Further results from the survey of adult skills*. Retrieved from http://www.keepeek.com/Digital-Asset-Management/oecd/education/skills-matter_9789264258051-en#.V-ZEjpMrKL4.

Phillips, D. (2006). Investigating policy attraction in education. *Oxford Review of Education, 32,* 551–559.

Robinson, K., & Aronica, L. (2015). *Creative schools: The grassroots revolution that's transforming education*. New York, NY: Viking.

Shibata, M. (2005). *Japan and Germany under the US occupation: A comparative analysis of post-war education reform*. Lexington books.

Shin, J. C. (2012). Higher education development in Korea: Western university ideas, Confucian tradition, and economic development. *Higher Education, 64*(1), 59–72.

Shook, J. (2010). How to change a culture: Lessons from NUMMI. *MIT Sloan Management Review, 51*(2), 63–68.

Tan, C. (2013). *Learning from Shanghai lessons on achieving educational success*. Dordrecht, Netherlands: Springer.

Terzian, S. G. (2008). Adventures in science: Casting scientifically talented youth as national resources on American radio, 1942–1958. *Paedagogica Historica, 44,* 309–325.

Tsui, A. (2007). Language policy and the construction of identity: The case of Hong Kong. In A. Tsui & J. Tollefson (Eds.), *Language policy, culture, and identity in Asian contexts* (pp. 121–141). Mahwah, NJ: Lawrence Erlbaum.

Tu, W. M. (1993). *Way, learning and politics: Essays on the Confucian intellectual*. Albany, NY: State University of New York Press.

Tu, W. M. (Ed.). (1996). *Confucian traditions in East Asian modernity: Moral education and economic culture in Japan and the four mini-dragons*. Boston, MA: Harvard University Press.

Tucker, M. S. (2011). *Surpassing Shanghai: An agenda for American built on the world's leading systems*. Boston, MA: Harvard Education Press.

Tweed, R. G., & Lehman, D. R. (2002). Learning considered within a cultural context: Confucian and Socratic approaches. *American Psychologist, 57*(2), 89–99.

Urban, W. J. (2010). *More than science and Sputnik: the National Defense Education Act of 1958*. Tuscaloosa, AL: University of Alabama Press.

Vogel, E. (1999). *Japan as number one: Lessons for America*. New York, NY: iUniverse.

Weale, S. (2015, November 26). Shanghai teaching method 'could improve UK results within four years. *The Guardian*. Retrieved from https://www.theguardian.com/education/2015/nov/26/shanghai-teaching-method-could-improve-uk-results-within-four-years.

Zuckerman, H. (1977). *Scientific elite*. New York, NY: The Free Press.

The manufacturer's authorised representative in the EU is Springer Nature Customer Service Centre GmbH, Europaplatz 3, 69115 Heidelberg, Germany. If you have any concerns regarding our products, please contact ProductSafety@springernature.com

Printed and bound by CPI Group (UK) Ltd, Croydon, CR0 4YY
30/09/2026
02225513-0015